Testing Statistical Assumptions in Research

Dedicated to
My wife Haripriya children Prachi-Ashish and Priyam,
– J. P. Verma

My wife, sweet children, parents, all my family and colleagues.
– Abdel-Salam G. Abdel-Salam

Testing Statistical Assumptions in Research

J. P. Verma
Lakshmibai National Institute of Physical Education
Gwalior, India

Abdel-Salam G. Abdel-Salam
Qatar University
Doha, Qatar

This edition first published 2019

© 2019 John Wiley & Sons, Inc.

Registered Offices
John Wiley & Sons, Inc., 111 River Street, Hoboken, NJ 07030, USA

Editorial Office
111 River Street, Hoboken, NJ 07030, USA

For details of our global editorial offices, customer services, and more information about Wiley products visit us at www.wiley.com.

Wiley also publishes its books in a variety of electronic formats and by print-on-demand. Some content that appears in standard print versions of this book may not be available in other formats.

Library of Congress Cataloging-in-Publication data applied for

Hardback ISBN: 9781119528418

Cover design: Wiley
Cover image: © JuSun/iStock.com

Set in 10/12pt WarnockPro by SPi Global, Chennai, India

Printed in the United States of America

V10008363_022119

Contents

Preface

The book titled *Testing Statistical Assumptions in Research* is a collaborative work of J. P. Verma and Abdel-Salam G. Abdel-Salam. While conducting research workshops and dealing with research scholars, we felt that most of the scholars do not bother much about the assumptions involved in using different statistical techniques in their research. Due to this reason, the results of the study become less reliable, and in certain cases, if the assumptions are severely violated, the results get completely reversed. In other words, in detecting the effect in hypothesis testing experiment, although the effect exists, it is not visible due to the extreme violation of assumptions. Since lots of resources and time are involved in conducting any survey or empirical research, one must test the related assumptions of statistical tests used in his or her analysis for making his or her findings more reliable.

The bringing out of this text has two specific purposes: first, we wish to educate the researchers about assumptions that are required to be fulfilled in different commonly used statistical tests and show the procedure of testing them using IBM SPSS®[1] Statistics software (SPSS) via illustrations. Second, we endeavor to motivate them to check assumptions by showing them the adverse effects of severely violating assumptions using specific examples. We have also suggested the remedial measures in using different statistical tests if assumptions are violated.

This book is meant for research scholars of different disciplines. Since most of the graduate students also write dissertations, this text is equally useful for them as well.

The book contains six chapters. Chapter 1 discusses the importance of assumptions in analyzing research data. The concept of hypothesis testing and statistical errors has been discussed in detail. We have also discussed the concept of power, sample size, and effect size and relationship among themselves in order to provide the foundation of hypothesis to the readers.

---

1 SPSS Inc. was acquired by IBM in October 2009.

In Chapter 2, we have introduced SPSS functionality to the readers. We have also shown how to segregate data, draw random samples, file split, and create variables automatically. These operations are extremely useful for the researchers in analyzing the survey data. Further, acquaintance with SPSS in this chapter shall facilitate the readers to understand the procedure involved in testing assumptions in different chapters.

In Chapter 3, we have discussed different assumptions required in survey studies. We have deliberated upon the importance of designing surveys in reporting the efficient findings.

Chapter 4 gives various parametric tests and the related assumptions. We have shown the procedures for testing these assumptions using the SPSS software. In order to motivate the readers to use assumptions, we have shown many situations where violation of assumptions affects the findings. In this chapter, we have discussed assumptions in all those statistical tests that are commonly used by researchers such as z, t, and F tests, ANOVA, correlation, and regression analysis.

In Chapter 5, we have discussed assumptions required for different nonparametric tests such as Chi-square, Mann-Whitney, Kruskal-Wallis, and Wilcoxon Signed-Rank test. We have shown the procedures of testing these assumptions as well.

Finally, in Chapter 6, assumptions in nonparametric correlations such as bi-serial correlation, tetrachoric correlation, and phi coefficient have been discussed. These types of analyses are often used by researchers in survey studies.

Hope this book will serve the purpose for which it has been written. The readers are requested to send their feedback about the book or about the problems that they encounter to the authors for further improvement in the text. You may contact the authors at their emails: Prof. J. P. Verma (email: vermajprakash@gmail.com, web: www.jpverma.org) and Dr. Abdel-Salam G. Abdel-Salam (email: abdelsalam811@gmail.com or abdo@vt.edu, web: https://www.abdo.website) for any help in relation to the text.

J. P. Verma
Abdel-Salam G. Abdel-Salam

Acknowledgments

We would like to thank our workshop participants, research scholars, and graduate students who have constantly posed innumerable problems during academic discussions, which have encouraged us to prepare this text. We extend our thanks to all those who directly or indirectly helped us in completing this text.

J. P. Verma
Abdel-Salam G. Abdel-Salam

About the Companion Website

This book is accompanied by a companion website:
www.wiley.com/go/Verma/Testing_Statistical_Assumptions_Research

The website includes the following:
- Chapter presentation in PPT format
- SPSS data file for each illustration where the data have been used.

1

Importance of Assumptions in Using Statistical Techniques

1.1 Introduction

All researches are conducted under certain assumptions. Validity and accuracy of findings depends upon whether we have fulfilled all the assumptions of data and statistical techniques used in the analysis. For instance, in drawing a sample, simple random sampling requires the population to be homogeneous while stratified sampling assumes it to be heterogeneous. In any research, certain research questions are framed that we try to answer by conducting the study. In solving these questions, we frame hypotheses that are tested with the help of the data generated in the study. These hypotheses are tested using some statistical tests, but these tests depend upon whether the data is nonmetric or metric. Different statistical tests are used for nonmetric and metric data for answering same research questions. More specifically, we use nonparametric tests for nonmetric data and parametric tests for metric data. Thus, it is essential for the researchers to understand the type of data generated in their studies. Parametric tests no doubt provide more accurate findings than the nonparametric tests, but they are based upon one common assumption of normality besides some specific assumptions associated with each test. If normality assumption is severely violated, the parametric tests may distort the findings. Thus, in research studies, assumptions are focused on two spheres: data and statistical tests besides methodological issues. Nowadays, many statistical packages such as IBM SPSS® Statistics software ("SPSS"),[1] Minitab, Statistica, and Statistical Analysis System (SAS) are available for analyzing both nonmetric and metric data, but they do not check the assumptions automatically. However, these software do provide outputs for testing associated assumptions with the statistical tests. We shall now discuss different types of data that can be generated in research studies. By knowing this, one can decide the relevant strategy for answering their research questions.

1 SPSS Inc. was acquired by IBM in October 2009.

Testing Statistical Assumptions in Research, First Edition. J. P. Verma and Abdel-Salam G. Abdel-Salam.
© 2019 John Wiley & Sons, Inc. Published 2019 by John Wiley & Sons, Inc.
Companion Website: www.wiley.com/go/Verma/Testing_Statistical_Assumptions_Research

1.2 Data Types

Data are classified into two categories: nonmetric and metric. Nonmetric data are also termed as qualitative and metric as quantitative. Nonmetric data are further classified as nominal and ordinal. Nonmetric data are a categorical measurement and are expressed by means of a natural language description. It is often known as "categorical" data. The data such as Student's Specialization = "Economics", Response = "Agree", Gender = "Male", etc. are examples of nonmetric data. These data can be measured on two different scales, i.e. nominal and ordinal.

1.2.1 Nonmetric Data

Nominal data are obtained by categorizing an individual or object into two or more categories, but these categories are not graded. For example, an individual can be classified into male or female category, but we cannot say whether male is higher or female is higher based on the frequency of the data set. Another example of nominal data is the color of the eye. One can be classified into blue, black, or brown eye categories. With this type of data, one can only compute percentage and proportion to know the characteristics of the data. Furthermore, mode is an appropriate measure of central tendency for such a data.

On the other hand, in the ordinal data, categories are graded. The order of items is often defined by assigning numbers to them to show their relative position. Here also, we classify a person, response, or object into one of the many categories, but we can rank them in some order. For example, variables that assess performance (excellent, very good, good, etc.) are ordinal variables. Similarly, attitude (agree, can't say, disagree) and nature (very good, good, bad, etc.) are also ordinal variables. On the basis of the order of an ordinal variable, one may not be sure as to which value is the best or worst on the measured phenomenon. Moreover, the distance between ordered categories is also not measurable. No mathematical operation can be done in the ordinal data. Median and quartile deviation are the appropriate measures of central tendency and variability, respectively, in such data.

1.2.2 Metric Data

Metric data are always associated with a scale measure, and therefore, it is also known as scale data. Such type of data are obtained by measuring some phenomena. Metric data can be measured on two different types of scale, i.e. *interval* and *ratio*. The data measured on interval and ratio scales are also termed as interval data and ratio data, respectively. Interval data are obtained by measuring a phenomenon along a scale where each position is equidistant from one another. In this scale, the distance between the two pairs are equivalent in some way. The only problem with this scale is that the doubling principle breaks down as there is no real zero on the scale. For instance, the eight marks given to an individual on the basis of his or her creativity do not explain that his or her creativity is twice as good as the person with four marks on a 10-point scale. Thus, variables measured on an interval scale have values in which differences are uniform and meaningful but ratios are not. Interval data may be obtained if the parameters such as motivation or level of adjustment is rated on a scale of 1–10.

The data measured on ratio scale has a meaningful zero and has an equidistant measure (i.e. the difference between 30 and 40 is the same as the difference between 60 and 70). Because zero exists in ratio data, 80 marks obtained by person A on a skill test may be considered twice the 40 marks obtained by another person B on the same test. In other words, doubling principle holds in ratio data. All types of mathematical operations can be performed with such kind of data. Examples of ratio data are weight, height, distance, salary, etc.

1.3 Assumptions About Type of Data

We know that for metric data, the parametric statistics are calculated while for nonmetric the nonparametric statistics are used. If we violate these assumptions, the findings may be misleading. We shall show this by means of an example. Before that let us elaborate data assumptions little more. If the data are nominal, we find mode as a suitable measure of central tendency, and if the data are ordinal, we compute median. Since both nominal and ordinal data are nonmetric, we use nonparametric statistics (mode and median). On the other hand, if the data are metric (interval/ratio), we should use parametric statistics such as mean and standard deviation. But we can calculate parametric statistics for the metric data only when the assumption of normality holds. In case the normality violates, we should use nonparametric statistics like median and quartile deviation. Assumptions of data in using measures of central tendency are summarized in Table 1.1.

Let us see what happens if we violate the assumption for the metric data. Consider the marks obtained by the students in an examination as shown in Table 1.2. This is a metric data; hence, without bothering about the normality assumption, let us compute the parametric statistic, mean. Here, the mean of the data set is 46. Can we say that the class average is 46 and report this finding in our research report? Certainly not, as most of the data are less than 46.

Let us see why this situation has arisen. If we look at the distribution of the data, it is skewed toward the positive side of the distribution as shown in Figure 1.1. Since the distribution of data is positively skewed, we can conclude that the normality assumption has been severely violated.

Table 1.1 Assumptions about data in computing measures of central tendency.

Data type	Nature of variable	Appropriate measure of central tendency
Nonmetric	Nominal data	Mode
	Ordinal data	Median
Metric	Interval/ratio (if symmetrical or nearly symmetrical)	Mean
	Interval/ratio (if skewed)	Median

Table 1.2 Marks for the students in an examination.

Student	1	2	3	4	5	6	7	8	9	10
Marks	35	40	30	32	35	39	33	32	91	93

Positively skewed distribution

Figure 1.1 Showing the distribution of data.

In a situation where the normality assumption is violated, we can very well use the nonparametric statistic such as median, as shown in Table 1.1. The median of this data set is 35, which can rightly be claimed as an average as most of the scores are around 35 in comparison to 46. Thus, if the data are skewed, then one should report median and quartile deviation as the measures of central tendency and variability, respectively, instead of mean and standard deviation in their project report.

1.4 Statistical Decisions in Hypothesis Testing Experiments

In hypotheses testing experiments, since population parameter is tested for some of its characteristics on the basis of the sample obtained from the population of interest, some errors are bound to happen. These errors are known as statistical errors. We shall investigate these errors and their repercussion in detail in the following sections.

1.4.1 Type I and Type II Errors

In hypotheses testing experiments, research hypothesis is tested by negating the null hypothesis. The focus of the researcher is to test whether the null hypothesis can be rejected on the basis of the given sampled data. The readers should note that a null hypothesis is never accepted. Either it is rejected or we fail to reject it on the basis of the given data. In hypothesis testing experiments, we test population characteristics on the basis of the sample; hence, some errors are bound to happen. Let us see what these errors are all about. While testing the null hypothesis, four types of decisions are possible, out of which two are correct and two are wrong. The two wrong decisions are rejecting the null hypothesis when it is true and failing to reject the null hypothesis when it is false. On the other hand, there are two correct decisions: rejecting the null hypothesis when it is not correct and not rejecting the null hypothesis when it is true. All these decisions have been summarized in Table 1.3.

The two wrong decisions discussed above are referred to as statistical errors. Rejecting a null hypothesis when it is true is known as "Type I error (α)" and failing to reject the null hypothesis when it is false is "Type II error (β)." Type I error is known as false positive because this error facilitates the researcher to accept the false claim. Similarly, Type II error is also known as false negative because this error guides the researcher not to accept the correct claim in the experiment. Since both the errors result in erroneous conclusion, the researcher always tries to minimize them. But the simultaneous minimization of both these errors, α and β, are not possible for a fixed sample size because if α decreases, then β increases and vice versa. If we wish to decrease these two errors simultaneously, sample size needs to be increased. But if the sample size cannot be increased, then one should fix the most severe error to an acceptable low level in the experiment. Out of the two errors, Type I error is more severe than Type II error. It is because Type I error forces the researcher to reject the correct null hypothesis and accept the false claim in the experiment. On the other hand, Type II error dictates the researcher not to accept the correct claim by not rejecting the null hypothesis. Since accepting the wrong claim

Table 1.3 Statistical errors in hypothesis testing experiment.

		Actual state	
		H_0 true	H_0 false
Researcher's decision	Reject H_0	Type I error (α) (false positive)	Correct decision ($1 - \beta$)
	Do not reject H_0	Correct decision	Type II error (β) (false negative)

Table 1.4 Implication of errors in hypothesis testing experiment.

		Actual state	
		H_0 true	H_0 false
Researcher's decision	Reject H_0	Type I error (α) (wrongly concluding that drug is effective)	Correct decision $(1 - \beta)$
	Do not reject H_0	Correct decision	Type II error (β) (wrongly rejecting the effective drug)

(α) is more serious than not accepting the correct claim (β), α is kept at low level in comparison to β.

Let us consider an experiment to test the effectiveness of a drug. Here, the null hypothesis H_0 is that the drug is not effective, which is tested against the alternative hypothesis H_1 that the drug is effective. In committing Type I error, the researcher will wrongly reject the null hypothesis H_0 and will accept the wrong claim about the drug to be effective. In other words, wrongly rejecting the null hypothesis will guide the researcher to accept the wrong claim that may be serious in nature.

On the other hand, if Type II error is committed, the researcher does not reject the null hypothesis and the correct claim about the effectiveness of the drug is rejected. Implications of these errors have been shown in Table 1.4.

Type I error can also be considered as consumer's risk and Type II error as producer's risk. For the researcher, consumer's risk is more serious than the producer's risk; hence, these two errors must be decided in advance in the experiment.

Probability of committing Type I error is known as level of significance and is denoted by α. Similarly probability of Type II error is represented by β. Thus, we can write:

- α = Prob. (Rejecting the null hypothesis when H_0 is true)
- β = Prob. (Not rejecting the null hypothesis when H_1 is true)

Generally, Type I error (α) is taken as 0.05 or 0.01, whereas Type II error (β) is kept at 0.2 or less in the study. Since choice of α depends upon the severity of accepting the wrong claim, it may be selected even lesser than 0.01 depending upon the situation.

1.4.2 Understanding Power of Test

Power of a test is the probability of rejecting the null hypothesis when the research hypothesis is true. In other words, it is the probability of rejecting null hypothesis correctly. The power of a test is computed by $1 - \beta$. If power is fixed at 0.8 in the drug testing experiment, it simply indicates that the probability of correctly rejecting the null hypothesis is 0.80. In other words, if the null hypothesis is rejected 80% of the time, the claim about the drug would be true. In estimating the sample size in the experiment, one needs to decide in advance as to how much power one wishes to have in the experiment. Logically, one should fix a power of at least 0.8 in the experiment. This indicates that 80% of the trials should reject the null hypothesis correctly. If power of the test is kept at 50% or less, then there is no meaning in performing the test. Simply toss a coin and decide whether the drug is effective or not.

1.4.3 Relationship Between Type I and Type II Errors

We know that Type I and Type II errors are related to each other. If one error decreases, another increases. Let us see how change in Type I error affects Type II error. We shall discuss this relationship with the help of an example. Let us suppose that the weight x is normally distributed with unknown mean μ and standard deviation 2 kg. For testing the hypothesis $H_0 : \mu = 60$ against $H_1 : \mu > 60$ at 5% level with 40 samples, let us see what would be the power in the test. We know that under the null hypothesis, sample mean follows normal distribution. To find the distribution of $\bar{x}$ under H_1, the populations mean needs to be specified. Let us assume that we wish to test the above-mentioned null hypothesis $H_0 : \mu = 60$ against the alternative hypothesis $H_1 : \mu = 64$. Then under H_0 and H_1, the distribution of $\bar{x}$ shall be as shown in Figure 1.2.

Here, α is the probability of rejecting H_0 when it is true. The null hypothesis H_0 is rejected if the test statistic falls in the rejection region as shown by the dotted area in Figure 1.2. On the other hand, β is the probability of not rejecting H_0 when H_1 is true and this area is denoted by the shaded lines. The power $(1 - \beta)$ is the remaining area of the normal distribution when H_1 is true. You can see from Figure 1.2 that if α decreases, then β increases, and as a result, the power decreases. It can be noticed from the figure that the amount of decrease in α is not equal to the amount of increase in β. It is interesting to note that on decreasing the probability of Type I error (α), power of the test $(1 - \beta)$ also decreases and vice versa.

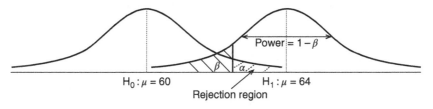

Figure 1.2 Distribution of mean under null and alternative hypotheses.

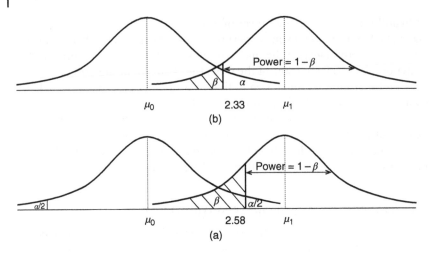

Figure 1.3 Showing comparison of power in (b) one- and (a) two-tailed tests at 1% level.

Often researchers feel elated if their null hypothesis is rejected at a significance level of 0.01 instead of 0.05 because they feel that their results are more powerful. If that is the case, why not reduce α to zero. In that case, the null hypothesis will never be rejected whatsoever the claim may be. On the other hand, if β is taken as 0, then every time null hypothesis would be rejected in favor of the claim. Thus, the researcher should choose α and β judiciously in the study.

1.4.4 One-Tailed and Two-Tailed Tests

In determining the sample size in hypothesis testing experiments, one needs to decide the type of test (one-tailed or two-tailed) to be used in the study. In a one-tailed test, the entire critical region, α, lies in one tail, whereas in a two-tailed test, it is divided to both the tails. Due to this reason, in a one-tailed test, β becomes less in comparison to that of a two-tailed test for the same α; this enhances power $(1 - \beta)$ in the one-tailed test. Thus, the power in the experiment can be increased at the cost of increasing α. The relationship between α and power $(1 - \beta)$ can be seen graphically in Figure 1.3. We may conclude that a one-tailed test is more powerful than a two-tailed test for the same level of α.

1.5 Sample Size in Research Studies

Findings of any research are based on the assumption that the appropriate number of sample elements has been selected in the study. The researchers are always in dilemma as to how much sample is enough for the study. A small sample provides inaccurate findings, whereas a large sample is an unnecessary use of extra resources like time and money. A small sample may not detect the effect in the experiment, whereas in a large sample, even a small effect may be significant, which may not have any practical utility. Criteria for choosing appropriate sample size depend upon whether the sample is being selected for the survey studies or for the hypotheses testing experiments. In survey studies, the sample size is determined on the basis of the amount of precision required in estimating a population parameter. On the other hand, in hypotheses testing experiments, it is estimated on the basis of the power required in the experiment. Details on sample size determination using software in research studies can be seen in the book titled *Determination of Sample Size and Power Analysis using G*Power Software* (Verma 2017).

To determine the sample size in survey studies, we need to know the population variability, amount of precision, and confidence coefficient required in estimating the population parameter. The following formula can be used to estimate the sample size:

$$n = Z_{\alpha/2}^2 \frac{pq}{d^2}$$

where p is the population proportion of the characteristics, generally estimated from similar studies conducted earlier, $q = 1 - p$, $Z_{\alpha/2}$ is the z value at a significance level α for a two-tailed test, and d is the precision required. Since p is rarely known, let us take p as 0.5 (assume maximum variance), and if we assume $Z_{\alpha/2}$ as approximately 2 (z value at 5% level), then sample size for different level of precisions at 5% level can be computed using the formula as shown in Table 1.5.

In a hypothesis testing experiment, a researcher often decides α and takes some sample data for testing his or her hypothesis on some population parameter. If the null hypothesis is rejected at say 5% level, then the conclusion is drawn that the treatment is effective. But the question is: How effective is

Table 1.5 Sample size for different precision.

Precision percentage	Precision level (d)	Sample size
10	0.1	100
5	0.05	400
4	0.04	625
3	0.03	1111
2	0.02	2500
1	0.01	10 000

the treatment? Simply by rejecting the null hypothesis, can we conclude that the treatment is effective in 95% trials if the same experiment is conducted 100 times? Here comes the role of defining power in the experiment. If the power is fixed at 0.9 in the experiment and the sample size is determined accordingly, then we can be sure that the null hypothesis will be correctly rejected 90% of the time.

Thus, for the researchers, two things are very important to fix in the hypothesis testing experiment: minimum effect that one wishes to test in the experiment and the power of the test besides Type I error at some predefined level. The whole discussion can be illustrated with an example from weight control research.

A health management company approached a big organization to sell its four-week weight management workshop to its obese employees. The company claims that its workshop will facilitate the employees to reduce their weight and improve functional efficiency. The CEO advised the company to conduct a research to see the effectiveness of the workshop on few employees, and if the findings of the workshop were encouraging, then he will allow the employees to join the workshop on a subsidized rate. During the experiment, the weight of the 15 participants were taken before and after the workshop. The null hypothesis in the experiment was that there will not be any change in the average weights of the participants against the research hypothesis that the post-workshop average weight will reduce significantly in comparison to the pre-workshop average weight. The null hypothesis was tested at 5% level. After analyzing the pre-post data on weight, the null hypothesis was rejected at 5% level. The company claimed that their workshop was effective at 5% level. In other words, 95% of the subjects reduced their weights. With this finding, should the CEO allow the company to launch the workshop for its employees? Several questions may be raised on the findings. First of all, how much weight reduction actually happened, and if 95% of the employees reduced their weights say only 150 g in four weeks, will the workshop be worth attending? Second question is: How much was the average weight of the subjects who participated in the workshop because a person weighing 100 kg will reduce faster than the person weighing 75 kg. The third question is: What was the power in the experiment? If the power of the test is only 0.5, then the rejection of the null hypothesis is not a guarantee of effectiveness of the program. In order to make the findings reliable, the researcher should have fixed the minimum detectable difference in post- and pre-workshop weight say 2 kg, the power say 0.9, and then the sample size required to detect 2-kg weight reduction should have been estimated. By freezing these conditions, if the null hypothesis would have been rejected in the experiment conducted on the estimated sample size, then one can say that the workshop will reduce the weight of the participants at least 2 kg in four-week time in 90% of the trials.

In hypothesis testing experiments, sample size is determined using the following information:

1) Minimum detectable difference (d)
2) Power in the test ($1 - \beta$)
3) Population variance (σ^2)
4) Type I error (α)
5) Type of the test (one-tailed or two-tailed)

While planning a hypothesis testing experiment, the researcher needs to decide and freeze the above-mentioned parameters to find the required sample size in the experiment. Using the following formula, the sample size can be computed. The meanings of the symbols are as usual:

$$n = \frac{\sigma^2}{d^2}(Z_\alpha + Z_\beta)^2$$

1.6 Effect of Violating Assumptions

In survey studies, we are mainly concerned with the precision and confidence coefficient in estimating population characteristics. Hence, violation of assumptions in such studies affects these two benchmarks. For instance, in case of extreme violation of normality assumption, the precision in estimating population parameter may not be same as claimed or the confidence coefficient in interval estimates may not be in conformity with what has been stated in the findings. Similarly, if the data is nonmetric and we apply parametric statistics for estimating population characteristics, the results will be completely absurd. Some of the assumptions may not be very serious but the others may affect the findings adversely. In case of extreme violation of certain assumptions, the results may be completely misleading. Thus, it is important for the researchers to investigate each and every assumption associated with the data and statistical techniques used in the survey studies carefully.

In hypothesis testing experiments, violation of assumptions for the data and statistical tests associated with the analysis affects three benchmarks: Type I error, power, and effect size. Depending upon the kind of violation of assumptions, these crucial factors get affected individually. The researcher may report the findings that the research hypothesis may be accepted at 5% significance level, but violation of assumption may actually raise the level to 10%. Similarly, if the study reports the power as 0.8, due to violation of assumption, it may actually be 0.6. Another disadvantage for not checking the assumptions may affect the claim on the effect size. In the weight control example discussed above, if the researcher claims that the minimum reduction of weight would be 2 kg if somebody participated in the workshop, this claim may be refuted in case of violation of the associated assumptions in the study.

What the researchers should do to avoid the pitfalls in reporting the findings? First, one should check the data type and choose the statistical analysis accordingly. If the data is metric, then one must check the normality assumption before using any parametric test. Second, one should investigate as to what are the associated assumptions of the selected statistical technique and check them accordingly.

Exercises

Multiple-Choice Questions

Note: Choose the most appropriate answer for each question.

1. Which of the following is affected by all the scores in a distribution?
   a. Mean
   b. Median
   c. Mode
   d. None of the above

2. If student's behavior has to be assessed by a panel of five judges, which of the following measures would be appropriate?
   a. Mean
   b. Median
   c. Mode
   d. None of the above

3. Which measures of central tendency will be suitable to compare height of the students in two classes?
   a. Mode
   b. Mean
   c. Median
   d. None

4. Which measure would be suitable to compare sale of different brands of mobile phones?
   a. Mean
   b. Median
   c. Mode
   d. Any of the above

5. Nominal data refers to the
   a. Metric data
   b. Continuous data
   c. Dichotomous data
   d. Integer data

6. Which of the following statement is not correct?
   a. Nominal data is also known as categorical data.
   b. In interval data, there is no real zero.
   c. In ratio data, the doubling principle holds.
   d. In ratio data, a person having scored 60 in a test may not be considered to be twice better than the one having scored 30.

7. Which of the following statement is correct?
   a. The mean can be computed for the nominal data.
   b. The mean is affected by the change of origin and scale.
   c. Mean can be computed for the ordinal data.
   d. Mean can be computed from truncated class interval.

8. Which of the following statement is not correct?
   a. The mean should not be calculated as a measure of central tendency if there are outliers in the data set.
   b. Median is not affected by the extreme scores.
   c. In skewed data, median is the best measure of central tendency.
   d. Outliers affect the mode.

9. Proportion and percentage statistics are best suited for which type of scores?
   a. Nominal
   b. Ordinal
   c. Interval
   d. Ratio

10. If there are no objective criteria of assessment, then the characteristics should be measured on which type of scale?
    a. Interval
    b. Ordinal
    c. Ratio
    d. Nominal

11. The Type I error can be best described by which of the following probability?
    a. P (Rejecting H_0/H_0 is true)
    b. P (Rejecting H_0/H_1 is true)
    c. P (Rejecting H_0/H_0 is false)
    d. P (Rejecting H_0/H_1 is false)

12. Which of the following probability defines Type II error correctly?
    a. P (Not rejecting H_0/H_1 is false)
    b. P (Not rejecting H_1/H_0 is false)
    c. P (Not rejecting H_0/H_1 is true)
    d. P (Not rejecting H_1/H_0 is true)

13. If the Type II error is represented by β, then power of the test is computed by
    a. $1 + \beta$
    b. $1 - \beta$
    c. $\beta - 1$
    d. β

14. The relationship between α and β can be best explained by
    a. $\alpha = \beta$
    b. $\alpha > \beta$
    c. $\alpha < \beta$
    d. $\alpha = k/\beta$

15. Choose the correct statement in relation to a research study.
    a. Type I and Type II errors are equally severe.
    b. Type I error is more severe than Type II error.
    c. Type II error is more severe than Type I error.
    d. Both the errors are not so severe.

16. The term $(1 - \alpha)$ can be defined as
    a. The probability of a Type I error
    b. The power of a test
    c. The probability of a Type II error
    d. The probability of not rejecting the null hypothesis when it is true

17. If a null hypothesis is not rejected at 5% level, what conclusion can be drawn?
    a. The null hypothesis will be rejected at 1% level of significance.
    b. The null hypothesis will be rejected at 10% level.
    c. The null hypothesis will not be rejected at 1% level.
    d. No conclusion could be drawn.

18. If the null hypothesis is rejected at 0.01 significance level, then
    a. It will also be rejected at the 0.05 significance level.
    b. It will not be rejected at the 0.05 significance level.
    c. It may not be rejected at the 0.05 significance level.
    d. It may not be rejected at the 0.02 significance level.

19. Other things being equal, which of the following actions will reduce the power in a hypothesis testing experiment?
    I. Increasing sample size
    II. Increasing significance level
    III. Increasing Type II error
    a. I only
    b. II only
    c. III only
    d. All of the above

20. Which of the following statement is not correct?
    a. Larger sample is required if Type I error decreases provided other conditions are same.
    b. Larger sample is required if Type II error increases provided other conditions are same.
    c. Sample size is directly proportional to the variability of the population.
    d. Sample size is inversely proportional to the effect size.

Short-Answer Questions

1. How many types of data exist in research? Explain them with examples. What kinds of analyses are possible with each data type?

2. Why Type I error is more serious in nature? Explain with examples.

3. What is the importance of power in hypothesis testing experiment? How can power be enhanced in testing?

4. Can both types of errors be reduced in research, if so how? Should Type I error be kept very low?

5. What are the various considerations in deciding the sample size in survey studies?

6. What are the considerations in deciding the sample size in hypothesis testing experiment?

Answers

Multiple-Choice Questions

1. a

2. b

3. b

4. c

5. c

6. d

7. b

8. d

9. a

10. b

11. a

12. c

13. b

14. d

15. b

16. d

17. c

18. a

19. c

20. b

2

Introduction of SPSS and Segregation of Data

2.1 Introduction

The kind of statistical test used for addressing the research questions depends upon the type of data generated in the study. Several statistical software are available for data analysis, but we have used the IBM SPSS® Statistics software ("SPSS")[1] in this book to show different analyses. "SPSS" stands for Statistical Package for Social Sciences. Using this software, different types of analyses can be carried out. While discussing data types in Chapter 1, we have seen that there are two different types of metric data, i.e. interval and ratio, but in SPSS, both these data types are referred to as Scale. In this chapter, we discuss the introduction of SPSS for the beginners along with different types of data management procedures that can be carried out in SPSS.

2.2 Introduction to SPSS

In this book, since we have used SPSS software for solving problems in different illustrations, it is important for us to introduce the readers to SPSS and its basic functionality. After installing the SPSS software on computer, click on SPSS icon in its directory. This will take you to Figure 2.1.

Check the radio button "Type in data" if the data file is prepared for the first time, and if an existing data file is to be used for the analysis, then check the radio button "Open an existing data source." Click **OK** to get Figure 2.2 for preparing data file. We shall discuss the procedure using the data shown in Illustration 2.1.

---

1 SPSS Inc. was acquired by IBM in October 2009.

Testing Statistical Assumptions in Research, First Edition. J. P. Verma and Abdel-Salam G. Abdel-Salam.
© 2019 John Wiley & Sons, Inc. Published 2019 by John Wiley & Sons, Inc.
Companion Website: www.wiley.com/go/Verma/Testing_Statistical_Assumptions_Research

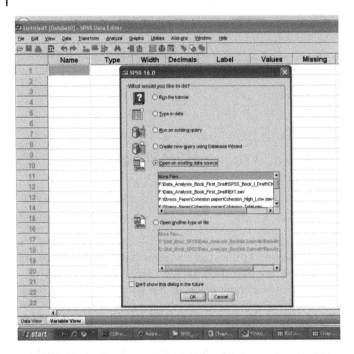

Figure 2.1 Screen for creating/opening data file. Source: Reprint Courtesy of International Business Machines Corporation, © International Business Machines Corporation.

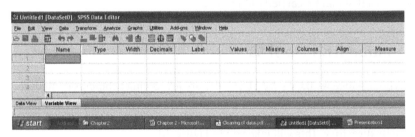

Figure 2.2 Screen for defining variables and their characteristics. Source: Reprint Courtesy of International Business Machines Corporation, © International Business Machines Corporation.

Illustration 2.1   The data shown in Table 2.1 represents the response of the subjects on the issue "Should Yoga be a compulsory activity in school?" Let us see how the data file in SPSS is prepared. The same data file shall be used for other analyses in this chapter.

Data file preparation is a two-step process. In the first step, we define all the variables and their characteristics, and in the second step, we feed the data for each variable.

Table 2.1 Response of the subjects and their demographic data on the issue "Should Yoga be a compulsory activity in school?"

ID	Age	Response	Gender	Height	District
1	23	Agree	Male	165	Delhi
2	25	Disagree	Male	167	London
3	43	Strongly agree	Female	164	Delhi
4	28	Disagree	Male	135	Sofia
5	19	Can't say	Female	169	Delhi
6	32	Agree	Female	171	Sofia
7	27	Agree	Male	172	London
8	24	Strongly disagree	Female	169	London
9	31	Can't say	Female	175	Sofia
10	25	Strongly agree	Male	168	Delhi

2.2.1 Data File Preparation

Step 1: Click on **Variable View** in Figure 2.2 for defining the variables in the SPSS. Using the following guidelines, all the variables and their characteristics can be defined.

Column 1: In the "Name" column, short names of the variables are defined. The variable names can have alphabet and numbers only, but the first letter of the name should essentially be an alphabet. The name of the variable cannot have two words. But if at all it is necessary, then they should be joined by underscore. For example, two-word names can be defined as Student_name, Blood_pressure, Chest_circumference, etc. One can write up to 64 bytes, which means 64 English letters in single-byte language.

Column 2: In the "Type" column, variable's type needs to be defined. Variables can be numeric or string. If the variables have comma or dot in between, then their type can be defined as "Comma" or "Dot," respectively. Similarly, variables can be defined of the type as "Scientific notation," "Date," "Dollar," etc. Variable's type can be defined by double clicking the cell.

Column 3: In the column "Width," length of variable's score is defined.

Column 4: In the "Decimal" column, the number of decimals a variable can have is defined.

Column 5: In the "Label" column, one can define the detailed name of the variable. Here, there is no restriction on using any type of alphabet, special character, or number in the variable name. Variable names can have more than one word. One can write up to 256 bytes, which means 256 English letters in single-byte language.

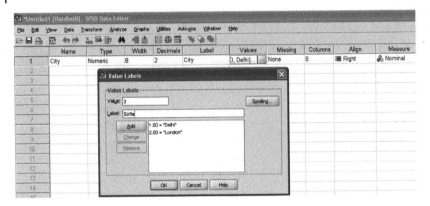

Figure 2.3 Defining code of nominal variable. Source: Reprint Courtesy of International Business Machines Corporation, © International Business Machines Corporation.

Column 6: In the column "Values," variable's code is defined by double clicking the cell. Coding of the variable is defined only for the nominal variable. For example, if there is a choice of choosing any one of the three states Delhi, London, and Sofia, then these district categories can be coded as 1 = Delhi, 2 = London, 3 = Sofia. SPSS window for defining the option for entering the code is shown in Figure 2.3.

Column 7: In survey studies, it is quite likely that a respondent may not reply certain questions. This creates the problem of missing value. Such missing values can be defined under the column heading "Missing." Data values that are specified as user-missing are flagged for some special treatment and are not included in most of the calculations.

Column 8: Under the heading "Columns," one can define the width of the column space, where data is typed in Data View.

Column 9: In the "Align" column, the type of alignment data should have (left, right, or center) can be defined.

Column 10: In the column "Measure," scale of measurement needs to be defined. In SPSS, interval and ratio data are defined as "Scale," whereas nominal and ordinal data are defined as "Nominal" and "Ordinal," respectively. After following the above-mentioned steps for defining variables in Table 2.1, the screen will look like Figure 2.4.

Remark To begin with, one can enter information only in three columns "Name," "Label," and "Measure" for Scale data, and for nominal data, one should define coding under column "Values" as well. For the rest of the columns, the default values are selected automatically.

Step 2: After defining all the variables and their properties in the variable view, click **Data View** in Figure 2.4 to get the screen for entering data column wise for all the variables. After entering the data, the screen will look like Figure 2.5.

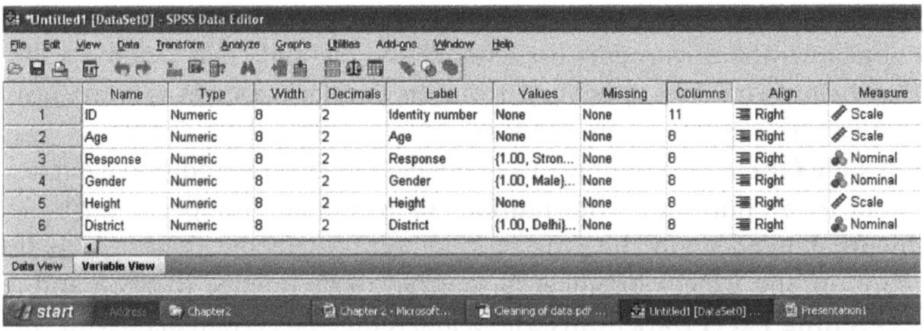

Figure 2.4 Screen after defining all the variables and their properties. Source: Reprint Courtesy of International Business Machines Corporation, © International Business Machines Corporation.

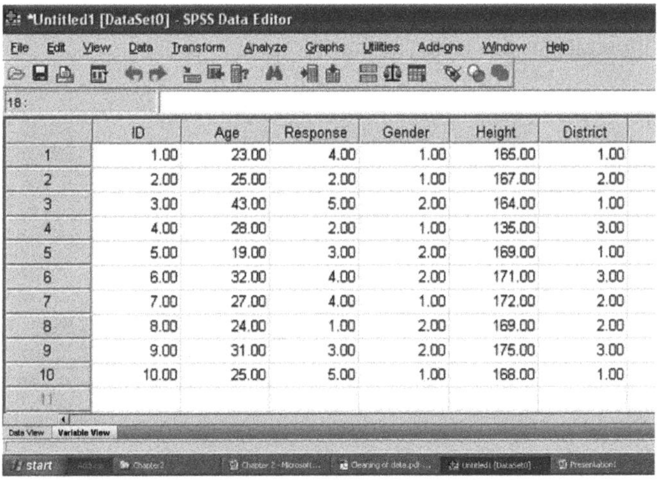

Figure 2.5 Screen after entering the data in SPSS file. Source: Reprint Courtesy of International Business Machines Corporation, © International Business Machines Corporation.

Remark Coding for different variables has been defined as follows:

Response 1: Strongly Disagree   2: Disagree   3: Can't Say   4: Agree   5: Strongly Agree

Gender 1: Male, 2: Female   **District:**   1: Delhi   2: London   3: Sofia

2.2.2 Importing the Data Set from Excel

The data set can either be imported from an existing file (e.g. Excel Worksheet) or be directly entered in the **Data View**. To import a data file from another source, follow the path *File → Open → Data*; the **Open Data** window will appear. Browse for the folder to locate the file to be opened. To open the file, choose the option **All Files** from the **Files of type** drop-down menu. The file will appear in the documents list, choose the file and click **Open**. In the illustration below, the data is opened from an Excel worksheet (Figure 2.6); however, it can be imported from other sources as well as the text files.

After clicking **Open**, the window(s) will open enquiring for additional details for the data to be imported. The number of windows to appear and their requirements depend upon the type of data source selected (Figure 2.7). In our illustration, the Excel worksheet is selected to be opened in SPSS and a window **Opening Excel Data Source** appears, requiring certain details for the data to be imported (variable names and data range).

Once the data file is ready, it should be saved in a desired location for performing different types of statistical analysis available under the **Analyze** command shown in the header. Two different types of data formats are generally used in SPSS depending upon the type of analysis we perform. These formats have been shown in relevant illustrations discussed in this book.

Figure 2.6 Data importing. Source: Reprint Courtesy of International Business Machines Corporation, © International Business Machines Corporation.

Figure 2.7 Data import options. Source: Reprint Courtesy of International Business Machines Corporation, © International Business Machines Corporation.

2.3 Data Cleaning

Data cleaning process is an important step in any research as it significantly affects the findings. It is the first step for every researcher in his or her study. One should check for the consistency of data and decide the strategy for missing data if any. These jobs can be done using the SPSS. In consistency check, we identify those data that are out of range. It may be due to the error in data entry or due to the extreme values. The missing response needs to be carefully handled to avoid the adverse effect in the results. Usually missing values are replaced by the group average or handled by case-wise deletion. Missing responses may create problem if their proportion to the total is more than 10%. We shall discuss the consistency check using the data shown in Table 2.1. In the data file shown in Figure 2.5, click **Data View** and press the sequence of commands **Analyze → Descriptive Statistics → Frequencies** as shown in Figure 2.8.

This will take you to Figure 2.9 to select the variables for analysis. Let us transfer all the variables from the left panel into right panel and press on **Statistics** to get Figure 2.10 to select the statistics, which are required to be computed by the SPSS. In this screen, let us select the option for mean, median, standard deviation, variance, minimum, maximum, skewness, and kurtosis. Click on **Continue** to get back to the screen in Figure 2.9.

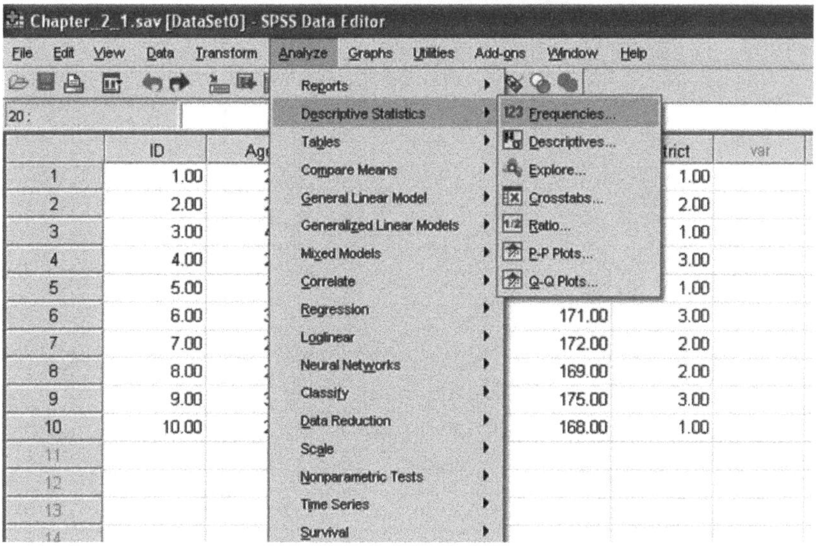

Figure 2.8 Screen showing command sequence for descriptive statistics. Source: Reprint Courtesy of International Business Machines Corporation, © International Business Machines Corporation.

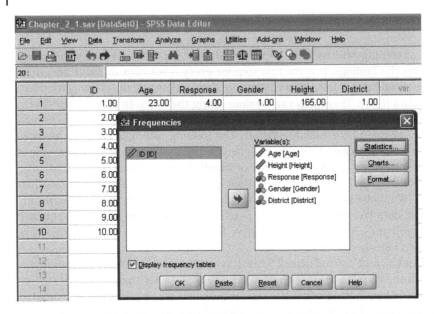

Figure 2.9 Screen showing selection of variables in the analysis. Source: Reprint Courtesy of International Business Machines Corporation, © International Business Machines Corporation.

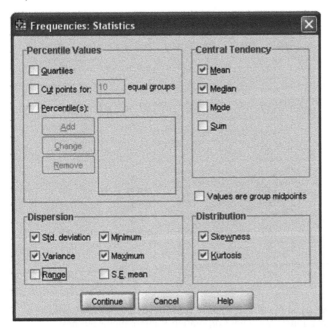

Figure 2.10 Screen for selecting different statistics for computation. Source: Reprint Courtesy of International Business Machines Corporation, © International Business Machines Corporation.

Figure 2.11 Screen for selecting chart type. Source: Reprint Courtesy of International Business Machines Corporation, © International Business Machines Corporation.

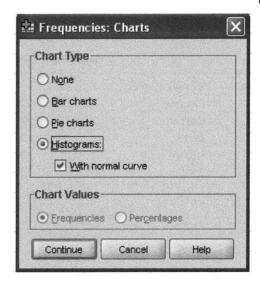

Now click on **Charts** to get Figure 2.11. In this screen, select option for "Histogram with normal curve." This will provide visual display of normality of different variables. Finally, pressing on **Continue** and **OK** on the screens will provide the outputs of descriptive statistics, frequency table, and histograms with normality graph for each variable. We have picked the frequency table and graphic for the "Response" variable only. These outputs are shown in Tables 2.2 and 2.3 and Figure 2.12.

Table 2.2 Descriptive statistics.

		Age	Height	Response	Gender	District
N	Valid	10	10	10	10	10
	Missing	0	0	0	0	0
Mean		27.70	1.66	3.30	1.50	1.90
Median		26.00	1.69	3.50	1.50	2.00
Std. deviation		6.59	1.12	1.34	0.53	0.88
Variance		43.34	125.39	1.79	0.28	0.77
Skewness		1.38	−2.67	−0.34	0.00	0.22
Std. error of skewness		0.69	0.69	0.69	0.69	0.69
Kurtosis		2.836	7.83	−0.85	−2.57	−1.73
Std. error of kurtosis		1.33	1.33	1.33	1.33	1.33
Minimum		19.00	135.00	1.00	1.00	1.00
Maximum		43.00	175.00	5.00	2.00	3.00

Table 2.3 Frequencies of "Response" variable.

		Frequency	Percent	Valid percent	Cumulative percent
Valid	Strongly disagree	1	10.0	10.0	10.0
	Disagree	2	20.0	20.0	30.0
	Can't say	2	20.0	20.0	50.0
	Agree	3	30.0	30.0	80.0
	Strongly agree	2	20.0	20.0	100.0
	Total	10	100.0	100.0	

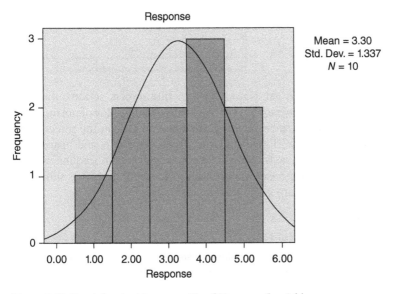

Figure 2.12 Graph for checking normality of "Response" variable.

2.3.1 Interpreting Descriptive Statistics Output

The first row in Table 2.2 represents name of each variable. The second row indicates the number of data for each variable ($n = 10$). The third row indicates whether any missing value exists or not.

The last two rows show the minimum and maximum for each data. For the entered data in SPSS, one should check whether minimum and maximum values for each of the variables are within the range. For instance, here subject's response was recorded on the issue "Should Yoga be a compulsory activity in school?" on a 5-point Likert scale. Thus, the minimum and maximum values of the "Response" variable should not be outside the limits of 1–5. On the other hand, if the values are outside this range, we should expect that the data has been entered incorrectly.

The skewness and kurtosis values of these variables give an idea about the distribution of scores. If skewness of any variable is more than twice its standard error, the data is considered as skewed, whereas the sign of skewness decides whether the data is positively or negatively skewed. Thus, we can see that the data on height is skewed because its skewness value (2.67) is greater than twice its standard error (2×0.69). Further, since the sign of the skewness is negative, we can interpret that the height data is negatively distributed. Thus, we may conclude that the height is not normally distributed.

2.3.2 Interpreting Frequency Statistic Output

For each of the variable, the SPSS shall generate frequency table along with its histogram with a normal distribution overplayed. Variable name is written in the heading as shown in Table 2.3. In the first column, value labels of the variable are shown. The second column indicates the frequency of responses that corresponds with each value label. The third, fourth, and fifth columns refer to the frequency. The third column indicates the percentage of each response, whereas the fourth column takes into account the missing data. If there is no missing data, percentage in the third and fourth column would be same. In the fifth column, cumulative percentage are shown, which adds up the percentage cumulatively.

Figure 2.12 shows the histogram of "Response" variable with a normal distribution overplayed. This gives the visual display of the data for checking normality graphically. This output also generates the mean, standard deviation, and number of subjects in the study. Using the frequency information, we can find out if there is any missing data and the extent of missing data. One can also check whether "value labels" have been entered correctly. By looking at the figure, we can get the feel about the normality of the data.

2.4 Data Management

After preparing data file in SPSS, several kinds of operations may be necessary in reorganizing the data as per requirements of the analysis. We shall discuss some of the most frequently used activities in data management with SPSS files. We shall restrict our discussion to sorting data, selecting cases, drawing random sample, splitting file, and computing variable. We shall discuss these operations using the data shown in Table 2.1.

2.4.1 Sorting Data

By sorting, we can reorganize the data of any variable in ascending or descending order. In survey studies, if the data has been entered at different point of time without organizing in a specific sequence, the sorting of data can provide some meaningful findings and allow us to use some specific applications in a better way. There are two options for sorting data: Sort Cases and Sort Variables.

2.4.1.1 Sort Cases

Using the "Sort Cases" option, one can rearrange the rows on a given variable(s) in the data file. Sorting in this way is also known as Row Sort. The entire rows can be sorted on selected variable(s) in ascending or descending order. After sorting cases, it is not possible to unsort the data to its original form. Thus, if the original order of rows is important, save the sorted file with different names.

Cases can be sorted using the sequence of commands **Data → Sort Cases** as shown in Figure 2.13. After pressing the commands, Figure 2.14 is obtained for selecting variables. Shift the variables "Response" and "District" from the left panel to the right panel. This will sort cases first by response and then by district. We have selected the radio button "Ascending" because we wish to sort cases in ascending order on response and district variables. Click on **OK** to get the output as shown in Figure 2.15, which contains cases sorted row wise on response and then on district.

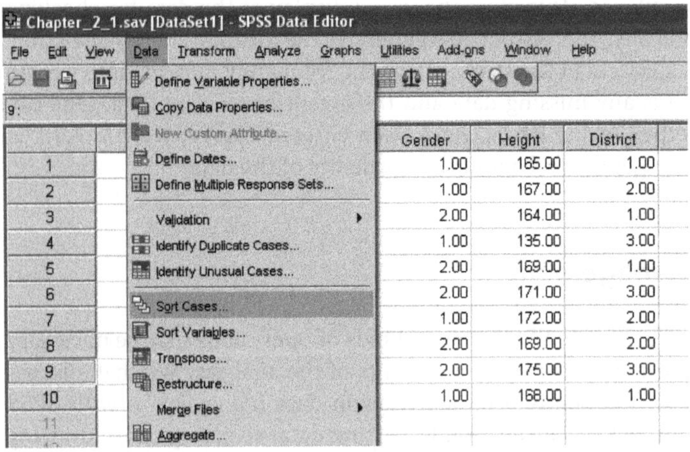

Figure 2.13 Screen showing commands for sorting cases. Source: Reprint Courtesy of International Business Machines Corporation, © International Business Machines Corporation.

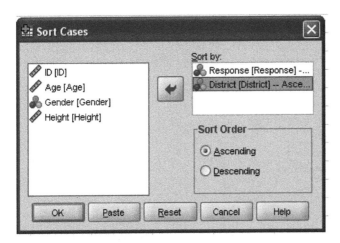

Figure 2.14 Screen for selecting variable(s) for sorting cases. Source: Reprint Courtesy of International Business Machines Corporation, © International Business Machines Corporation.

	ID	Age	Response	Gender	Height	District
1	8.00	24.00	1.00	2.00	169.00	2.00
2	2.00	25.00	2.00	1.00	167.00	2.00
3	4.00	28.00	2.00	1.00	135.00	3.00
4	5.00	19.00	3.00	2.00	169.00	1.00
5	9.00	31.00	3.00	2.00	175.00	3.00
6	1.00	23.00	4.00	1.00	165.00	1.00
7	7.00	27.00	4.00	1.00	172.00	2.00
8	6.00	32.00	4.00	2.00	171.00	3.00
9	3.00	43.00	5.00	2.00	164.00	1.00
10	10.00	25.00	5.00	1.00	168.00	1.00

Figure 2.15 Data file with sorted cases on response and district variables. Source: Reprint Courtesy of International Business Machines Corporation, © International Business Machines Corporation.

2.4.1.2 Sort Variables

This type of sorting is also known as column sort. In this type of sorting, variables are sorted column wise on the basis of the name, type, decimal, width, etc. In **Data View,** click the command sequence **Data → Sort Variables** as shown in Figure 2.16. This will take you to Figure 2.17.

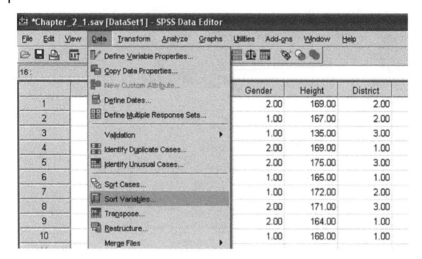

Figure 2.16 Screen showing commands for sorting variables. Source: Reprint Courtesy of International Business Machines Corporation, © International Business Machines Corporation.

Figure 2.17 Screening showing option for sorting criterion. Source: Reprint Courtesy of International Business Machines Corporation, © International Business Machines Corporation.

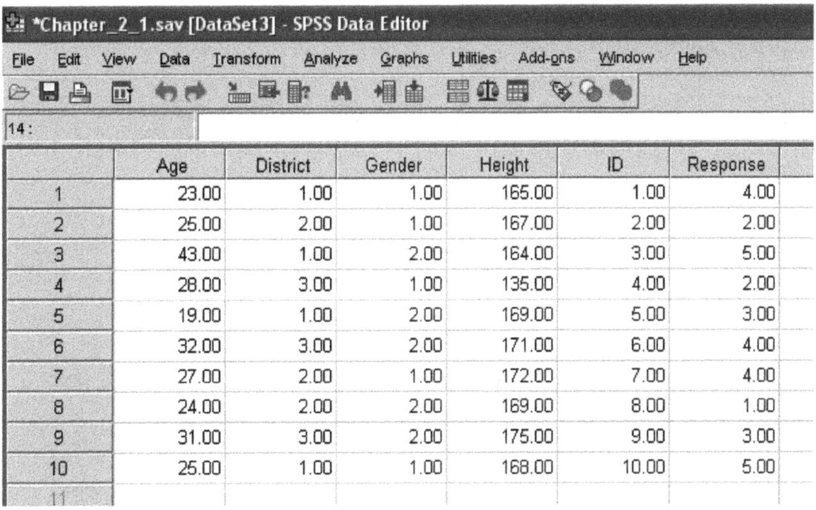

	Age	District	Gender	Height	ID	Response
1	23.00	1.00	1.00	165.00	1.00	4.00
2	25.00	2.00	1.00	167.00	2.00	2.00
3	43.00	1.00	2.00	164.00	3.00	5.00
4	28.00	3.00	1.00	135.00	4.00	2.00
5	19.00	1.00	2.00	169.00	5.00	3.00
6	32.00	3.00	2.00	171.00	6.00	4.00
7	27.00	2.00	1.00	172.00	7.00	4.00
8	24.00	2.00	2.00	169.00	8.00	1.00
9	31.00	3.00	2.00	175.00	9.00	3.00
10	25.00	1.00	1.00	168.00	10.00	5.00
11						

Figure 2.18 Screening showing sorted data column wise based on variable's name. Source: Reprint Courtesy of International Business Machines Corporation, © International Business Machines Corporation.

By selecting any of the option, the variables will be sorted accordingly. For instance, if we choose "Name," then the data column will be sorted on the basis of variable's name alphabetically. Similarly, if the sorting is done on "width," then the data columns will be sorted on the basis of variable's width defined in the data file. After selecting options "Name" and radio button "Ascending" in Figure 2.17 and by pressing **OK**, we shall get the output in sorted form on the basis of the variable's name as shown in Figure 2.18.

2.4.2 Selecting Cases Using Condition

In the database, cases can be selected using conditions. For instance, we can select only those cases that belong to the male or we may select the cases that belong to Delhi and have responded "Strongly Agree" to the issue "Yoga should be the compulsory activity in school" in the data set discussed above. Since all the categories of the nominal variables like District, Gender, and Response have been coded, the data in the SPSS will be shown in numerical form. However, if you want to see the categories you can do so just by pressing the icon of "Value Labels" in the header as shown in Figure 2.19. The data file after clicking on "Value Labels" icon in the header will look like as shown in Figure 2.20.

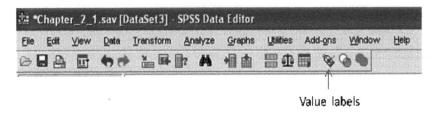

Value labels

Figure 2.19 Command for switching data from coding to categories and vice versa. Source: Reprint Courtesy of International Business Machines Corporation, © International Business Machines Corporation.

	ID	Age	Response	Gender	Height	District
1	1.00	23.00	Agree	Male	165.00	Delhi
2	2.00	25.00	Disagree	Male	167.00	London
3	3.00	43.00	Strongly Agree	Female	164.00	Delhi
4	4.00	28.00	Disagree	Male	135.00	Sofia
5	5.00	19.00	Can't Say	Female	169.00	Delhi
6	6.00	32.00	Agree	Female	171.00	Sofia
7	7.00	27.00	Agree	Male	172.00	London
8	8.00	24.00	Strongly Disagree	Female	169.00	London
9	9.00	31.00	Can't Say	Female	175.00	Sofia
10	10.00	25.00	Strongly Agree	Male	168.00	Delhi

Figure 2.20 Data file showing nominal variables as per their categories. Source: Reprint Courtesy of International Business Machines Corporation, © International Business Machines Corporation.

2.4.2.1 Selecting Data of Males with Agree Response

To select all those cases where "Male" subjects had "Agree" response, we shall use the command sequence **Data → Select Cases** as shown in Figure 2.21. This will take us to Figure 2.22.

In the first step, select the radio button attached to "If condition is satisfied" and press on **If** to get Figure 2.23 to define the required condition.

In the second step, write the condition **(Gender = 1) & (Response = 4)** for selecting cases if it is satisfied. While writing condition, bring the variables "Gender" and "Response" from the left panel to the blank space in the right panel using arrow key. It is advisable to use the mathematical symbols given on the screen rather than inserting from the keyboard. After defining the condition, press **Continue** to go back to Figure 2.22 for defining name of the file

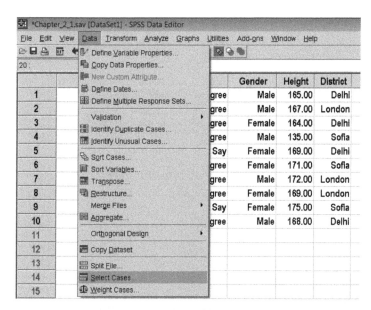

Figure 2.21 Screen showing commands for selecting cases. Source: Reprint Courtesy of International Business Machines Corporation, © International Business Machines Corporation.

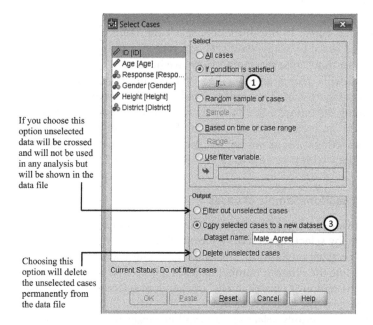

Figure 2.22 Screen for selecting cases. Source: Reprint Courtesy of International Business Machines Corporation, © International Business Machines Corporation.

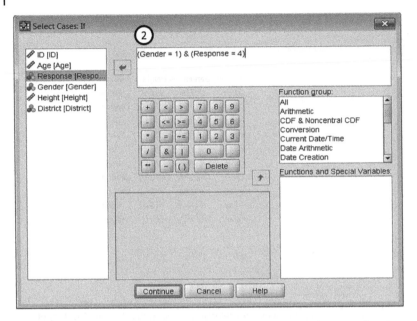

Figure 2.23 Screen showing condition for selecting cases with Male responding Agree. Source: Reprint Courtesy of International Business Machines Corporation, © International Business Machines Corporation.

containing selected cases. In the third step, select the radio button attached to "Copy selected cases to a new data set" and write the name of the file as you wish in the section "Data set name." Here, we have used the file name as Male_Agree. This will create a new data file with this name containing selected cases as shown in Figure 2.24. The readers should note that same restriction is applied in defining the name of the new data file as it was discussed in defining variable's name in creating the data file in SPSS. One can define different conditions for selecting cases as per the requirements. Some of the criteria for selecting cases along with the commands are shown in Table 2.4.

2.4.3 Drawing Random Sample of Cases

Using SPSS software, we can select random sample of cases from the data file. This can be done by selecting the radio button "Random sample of cases" in Figure 2.22 and pressing on **Sample**. This will create a new screen as shown in Figure 2.25 to define the random number of cases required to be selected.

If we choose the option "Approximately," our random selection of cases may not be exactly equal to the defined one due to its algorithm. But in case of large sample, number of randomly selected cases will be closer to the defined one. On the other hand, if we choose the second option "Exactly" and define the

	ID	Age	Response	Gender	Height	District
1	1.00	23.00	Agree	Male	165.00	Delhi
2	7.00	27.00	Agree	Male	172.00	London
3						

*Untitled3 [Male_Agree] - SPSS Data Editor

File Edit View Data Transform Analyze Graphs Utilities Add-ons Window Help

24:

Figure 2.24 Screen showing selected cases with Male having response Agree. Source: Reprint Courtesy of International Business Machines Corporation, © International Business Machines Corporation.

Table 2.4 Commands for selecting cases with different conditions in SPSS.

SN	Selection criteria	Conditions
1.	All data rows for male	Gender = 1
2.	Females having age less than 28 years	(Gender = 2) & (Age < 28)
3.	Males belonging to either Delhi or London	(Gender = 1) & ((District = 1) or (District = 2))
4.	All those responding either Agree or Strongly Agree	(Response = 4) or (Response = 5)

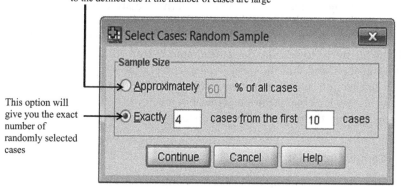

Choosing this option may not provide you the exact percentage of random cases due to algorithm used in small data file. However, the percentage of cases selected will be more closer to the defined one if the number of cases are large

This option will give you the exact number of randomly selected cases

Figure 2.25 Screen showing option for selecting specified number of cases randomly. Source: Reprint Courtesy of International Business Machines Corporation, © International Business Machines Corporation.

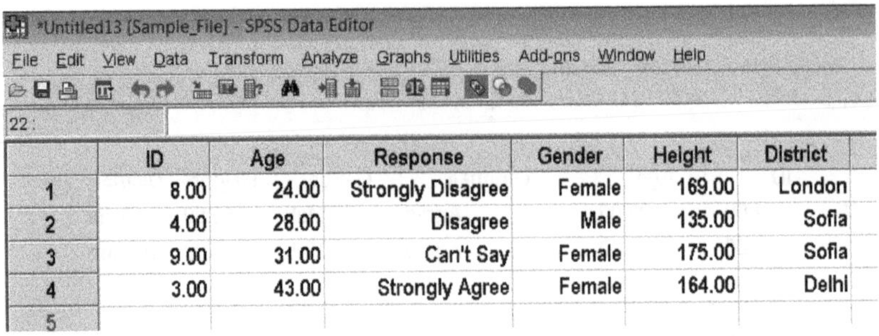

	ID	Age	Response	Gender	Height	District
1	8.00	24.00	Strongly Disagree	Female	169.00	London
2	4.00	28.00	Disagree	Male	135.00	Sofia
3	9.00	31.00	Can't Say	Female	175.00	Sofia
4	3.00	43.00	Strongly Agree	Female	164.00	Delhi
5						

Figure 2.26 Screen showing randomly selected cases. Source: Reprint Courtesy of International Business Machines Corporation, © International Business Machines Corporation.

number of cases required to be randomly selected from the total number of cases, this will provide the exact number of cases. In this case, we wish to select 4 randomly selected cases out of 10; hence, we have filled the entries accordingly in Figure 2.25. After pressing on **Continue**, we shall get back to Figure 2.22, in which after defining the file name, the reduced data set with randomly selected cases can be obtained as shown in Figure 2.26.

2.4.4 Splitting File

Sometimes we wish to analyze the data gender wise, district wise, or response wise. We can do so just by splitting the file on that variable. For instance, if we wish to find the mean and standard deviation of the subjects gender wise, one of the methods is to prepare one file for male and another for female using the procedure discussed above and then compute these statistics separately. But simply by splitting the file, we can compute all the statistics gender wise using the same data file. The file can be split using the command **Data → Split File** as shown in Figure 2.27. This will take us to Figure 2.28 for defining the variable on which the file needs to be split. Select the radio button attached to "Organize output by groups" and shift the variable Gender from the left panel to the right panel in the section "Group Based on." After pressing on **OK**, the file will be split. The file will look like the original one, but when you compute any statistic or do any analysis, the results will be produced gender wise.

2.4.5 Computing Variable

Sometimes, we need to compute a variable using two or more variables in the database. This can be done by defining the computing variable in SPSS. We shall discuss the procedure by means of Illustration 2.2.

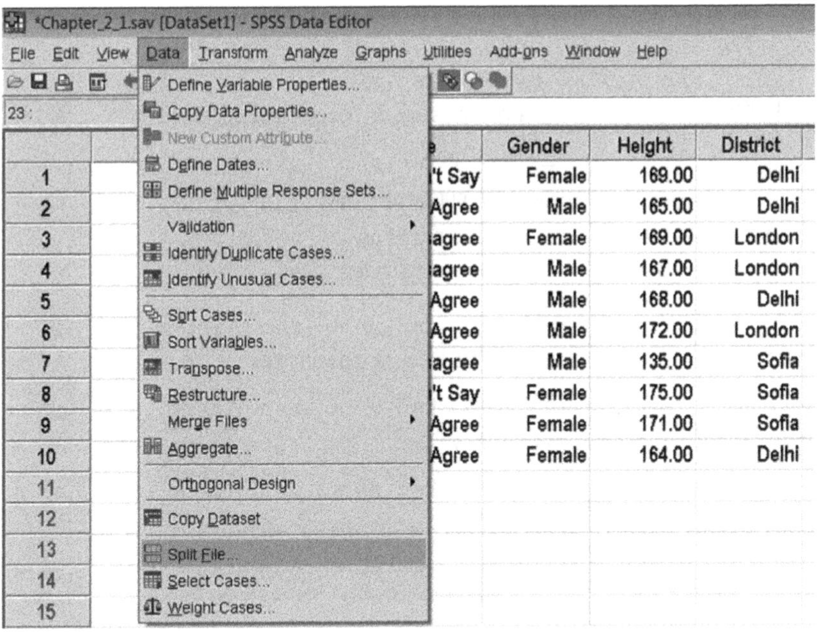

Figure 2.27 Screen showing commands for splitting the file. Source: Reprint Courtesy of International Business Machines Corporation, © International Business Machines Corporation.

Illustration 2.2 If the weight of the subjects is also included in the data file shown in Illustration 2.1, then let us see how the body mass index (BMI) can be calculated using the commands in SPSS.

Following a similar procedure as discussed in Illustration 2.1, the new data file can be prepared as shown in Figure 2.29. Here, we wish to compute BMI of all the subjects. One of the ways is to compute BMI for each individual separately and then enter in the SPSS data file for further analysis. But this job can be done using the commands **Transform → Compute Variable**. After pressing these commands in the header, we shall get Figure 2.30 for defining the formula.

Type the new variable name in the "Target Variable" section. Write the formula **Weight/(Height * Height)** in the "Numeric Expression" section in the right panel. While writing the formula, it is advisable to shift the variable's name from the left panel instead of typing and also use the mathematical commands from the screen instead of inserting from the keyboard. Once you ensure that the formula is correctly written, press **OK** to get the data file with newly created BMI variable as shown in Figure 2.31.

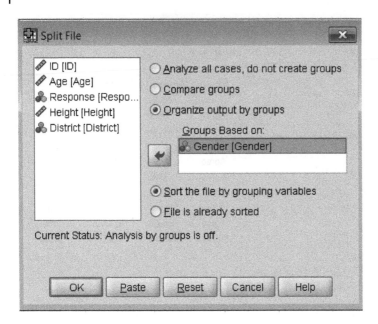

Figure 2.28 Screen showing option for defining the split variable. Source: Reprint Courtesy of International Business Machines Corporation, © International Business Machines Corporation.

	ID	Age	Response	Gender	Height	District	Weight
1	1.00	23.00	4.00	1.00	1.65	1.00	58.00
2	2.00	25.00	2.00	1.00	1.67	2.00	61.00
3	3.00	43.00	5.00	2.00	1.64	1.00	58.00
4	4.00	28.00	2.00	1.00	1.35	3.00	62.00
5	5.00	19.00	3.00	2.00	1.69	1.00	49.00
6	6.00	32.00	4.00	2.00	1.71	3.00	68.00
7	7.00	27.00	4.00	1.00	1.72	2.00	55.00
8	8.00	24.00	1.00	2.00	1.69	2.00	52.00
9	9.00	31.00	3.00	2.00	1.75	3.00	62.00
10	10.00	25.00	5.00	1.00	1.68	1.00	58.00

Figure 2.29 SPSS data file. Source: Reprint Courtesy of International Business Machines Corporation, © International Business Machines Corporation.

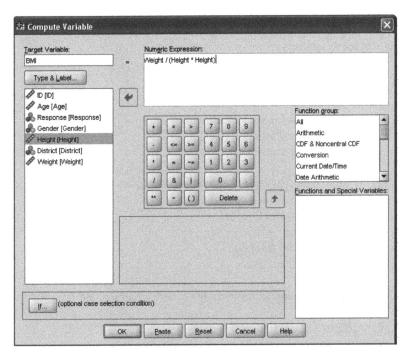

Figure 2.30 Screen for defining formula for computing variable. Source: Reprint Courtesy of International Business Machines Corporation, © International Business Machines Corporation.

	ID	Age	Response	Gender	Height	District	Weight	BMI
1	1.00	23.00	4.00	1.00	1.65	1.00	58.00	21.30
2	2.00	25.00	2.00	1.00	1.67	2.00	61.00	21.87
3	3.00	43.00	5.00	2.00	1.64	1.00	58.00	21.56
4	4.00	28.00	2.00	1.00	1.35	3.00	62.00	34.02
5	5.00	19.00	3.00	2.00	1.69	1.00	49.00	17.16
6	6.00	32.00	4.00	2.00	1.71	3.00	68.00	23.26
7	7.00	27.00	4.00	1.00	1.72	2.00	55.00	18.59
8	8.00	24.00	1.00	2.00	1.69	2.00	52.00	18.21
9	9.00	31.00	3.00	2.00	1.75	3.00	62.00	20.24
10	10.00	25.00	5.00	1.00	1.68	1.00	58.00	20.55
11								

Figure 2.31 Screen showing data file with newly created BMI variable. Source: Reprint Courtesy of International Business Machines Corporation, © International Business Machines Corporation.

Exercises

Multiple-Choice Questions

Note: Choose the best answer for each question.
1. SPSS stands for
    a. Statistical package for Social Studies
    b. Statistical package for Social Sciences
    c. Social Package for Statistical Science
    d. Social package for Statistical Studies

2. While defining data types in SPSS, which of the following nomenclatures are used
    a. Interval, Nominal, and Ordinal
    b. Nominal, ratio, and ordinal
    c. Nominal, Ratio, and Interval
    d. Nominal, Ordinal, and Scale

3. Which of the following statements is correct while making data file in SPSS?
    a. Variables can also be defined in data view.
    b. Variable's name can only be entered in variable view.
    c. In data view, besides entering data, variable's name can also be altered.
    d. Data can also be typed in variable view after defining the variables.

4. In preparing a new data file in SPSS, which option is used?
    a. Open another type of file
    b. Type in data
    c. Open an existing data source
    d. Run the tutorial

5. Which is the valid variable name in SPSS?
    a. MuscularStrength
    b. Muscular Strength
    c. Muscular-Strength
    d. "Muscular_Strength"

6. While defining the value labels for three categories of a nominal variable, which is the correct option?
    a. 1-Statistics, 2-Economics, 3-Military Science
    b. 0-Statistics, 1-Economics, 2-Military Science
    c. 2-Statistics, 4-Economics, 6-Military Science
    d. All are correct

7. While data cleaning, if the outliers are removed, what impact will it have?
   a. Data will result in accurate findings.
   b. Data will become random sample.
   c. Data may become normal.
   d. Data can be used for any type of analysis.

8. If data is skewed, what should be reported as measures of central tendency and variability, respectively?
   a. Mean and quartile deviation
   b. Median and standard deviation
   c. Mean and standard deviation
   d. Median and quartile deviation

9. If the data of height is negatively skewed, what conclusion can be drawn?
   a. Most of the people are taller than the average height of the group.
   b. Most of the people are shorter than the average height of the group.
   c. Equal number of people are on both sides of the average height of the group.
   d. No conclusion can be drawn.

10. Consistency check of the variable can be done using
    a. Skewness
    b. Kurtosis
    c. Minimum and maximum scores
    d. Missing data

11. If missing value exists in metric data, what should be the correct strategy?
    a. It should be replaced by the highest score in the data set.
    b. It should be replaced by the lowest scores in the data set.
    c. It should be replaced by the next immediate score.
    d. It should be replaced by the average score of the data set.

12. The most important interface to define properties of the variables in the data set is
    a. Data view
    b. Output window
    c. Variable view
    d. None of the above

13. Kurtosis is significant if
    a. Kurtosis > 2SE(Kurtosis)
    b. Kurtosis < 2SE(Kurtosis)
    c. Kurtosis > SE(Kurtosis)
    d. Kurtosis < SE(Kurtosis)

14. If Gender (male = 1 and female = 2) and City (Delhi = 1, London = 2, and Sofia = 3) have been defined as nominal variables, which command you would use to select all males living in Delhi?
    a. Gender, City = 1
    b. Gender = 1 & City = 1
    c. Gender = 1 or City = 1
    d. (Gender = 1)(City = 1)

15. What happens if the file is split on Gender in SPSS?
    a. Two files are created with different names.
    b. The data set is split into two parts but not visible and you need to specify gender-wise analysis for results.
    c. Two files are created automatically with names Male and Female.
    d. The analysis results are generated gender-wise automatically.

Short-Answer Questions

1. What are the essential steps in making a data file in SPSS?

2. Why data cleaning is important in research? Explain with example.

3. What is the significance of skewness and kurtosis in explaining research findings? Can an intervention be tested for its effectiveness using skewness and kurtosis?

4. Discuss the importance of percentile in explaining research findings.

5. Should you always present mean and standard deviation in the research report? Give your comments.

Answers

Multiple-Choice Questions

1. b

2. d

3. b

4. b

5. a

6. d

7. c

8. d

9. a

10. c

11. d

12. c

13. a

14. b

15. d

3

Assumptions in Survey Studies

3.1   Introduction

In survey studies, lots of resources and efforts are required in collecting and analyzing the data. After all the hard work, it is desired that the inferences are correct. We wish to ensure that any difference observed in data samples is actually due to difference in their population and not merely due to chance. Similarly, in estimating some population characteristics on the basis of survey data, we wish that the estimate is reliable. All this can be achieved if we fulfill the assumptions associated with survey methodology, questionnaire, data, and analysis tools used in the study.

Testing assumptions in survey research is the necessary step as they enable us to conduct the studies. Sometimes, some of the assumptions are difficult to prove, but we ensure them by indirect approach. For instance, we assume that the participants will give honest and correct response, but we cannot check this assumption directly. However, we ensure this by proper use of wordings in the questionnaire and effective communication of instructions to the respondents.

Even if some of the assumptions cannot be directly proved, we need to justify that none of the assumptions will be violated in the study; otherwise, the study will not be of any use. In each study, it is essential to explain about delimitation and limitation. Delimitation refers to the boundary conditions in terms of population, research questions, and parameters under study. These should be clearly defined. The researcher must explain categorically about the population of interest. The idea behind restricting population in terms of age, gender, region, etc. is to control variability. We saw in Chapter 2 that the population variability is one of the essential parameters in determining the sample size. Thus, if the study is not delimited, then the variability of the population would be large, which in turn requires larger sample, and at the same time, reliability of the findings will also be an issue. Similarly, limitations should also be clearly defined so as to facilitate the readers to view the findings

Testing Statistical Assumptions in Research, First Edition. J. P. Verma and Abdel-Salam G. Abdel-Salam.
© 2019 John Wiley & Sons, Inc. Published 2019 by John Wiley & Sons, Inc.
Companion Website: www.wiley.com/go/Verma/Testing_Statistical_Assumptions_Research

under the specified limitations. In this chapter, we shall discuss different types of assumptions associated with the survey procedures, questionnaire, data, and analysis that a researcher should understand.

3.2 Assumptions in Survey Research

While conducting survey, there are some specific guidelines that the researcher should follow in order to get the reliable information from the respondents. We shall discuss each one of them in this section.

3.2.1 Data Cleaning

Data cleaning is the first step of data analysis in any survey research. This ensures consistency of data obtained in the survey. We have seen the procedure for data cleaning in Chapter 2. Data cleaning is done for identifying outliers in the data set. This may exist due to wrong entry or due to extreme data in the set. In either case, it is important to identify such data because outliers might distort the entire findings at times. These outliers can also be identified using the boxplot. The detailed procedure has been discussed in Chapter 4. Another issue in data cleaning is the missing information of a subject on some parameters. This should also be tackled before proceeding for analysis as per the procedure discussed in Chapter 2. Thus, data cleaning ensures the detection of outliers if any, consistency check, and action about missing value. Cleaning data ensures valid results.

3.2.2 About Instructions in Questionnaire

It is our experience that the respondents hardly read the instructions in the questionnaire; hence, the questions should be framed in such a manner so that the message can be effectively conveyed to the respondents. For instance, opening note for few questions in the questionnaire may be written as "Following questions will decide the strategy for implementing the policy regarding the use of mobile at work place; hence, your views shall be very important." Such statements will not only motivate the respondents intrinsically but also facilitate the investigator to get authentic responses. There should be sufficient scope of dealing with some incomplete survey because in spite of all the efforts, it is difficult to ensure that all the respondents shall provide the complete information in the questionnaire. Thus, we assume that the respondents have understood the purpose of the survey correctly and the research design is sufficient to deal with some of the incomplete information.

3.2.3 Respondent's Willingness to Answer

To get the correct information, the respondents should be genuinely interested to take part in your survey. It is very important that they should be motivated by means of some incentives. If your respondents are from the higher socioeconomic group, intrinsic motivation may be good enough, but if they belong to the middle and low socioeconomic groups, then both intrinsic and extrinsic motivations should be ensured for them to be genuinely interested in the survey. While preparing the research proposal, some budget allocation must be kept as motivational cost. It's a good gesture to offer some small incentive to the respondents in the form of pen, pocket diary, calendar, etc., as per the budget allocation while asking them to join the survey. Another important feature for respondent's willingness to reply your questions is the time taken to complete the survey. Large number of questions will consume more time, which may deter the respondents from participating in your survey; hence, optimum number of questions that cover your entire subject matter should be included in the questionnaire.

3.2.4 Receiving Correct Information

One of the assumptions in survey research is that the respondents are giving correct information. If this is not the case, the whole purpose of the survey would be lost. Usually, most of the respondents do not read the instructions of the questionnaire seriously; hence, it is important that the researcher must explain the purpose of the study and also explain the significance of replying the questions correctly in the questionnaire. Respondents must be categorically informed that there is no correct and wrong answer and they need to reply as to what they feel. This ensures them not to attach their false ego in replying the questions. For instance, if the respondents are asked as to how much budget they allocate for sports and outdoor activities, they may overstate in replying such questions due to their self-image. Similarly, if the question about the salary is asked from the men or age from women, they are likely to give false information due to fear of social image. Another way to enhance the correctness of the respondent's information is not to ask them about their personal identity and confirm about the confidentiality of their information. This will ensure correct details about the salary, savings, expenditure, etc. of the respondents.

3.2.5 Seriousness of the Respondents

During the study, we need to assume that the respondents are serious in responding the questionnaire. Due to nonserious attitude of the respondents, the results may be totally misleading. Although consent of the respondents to take part in the survey study is taken, in spite of the willingness, one can never assume that all of them will take it seriously. Therefore, the survey should be idiot-proof. In other words, if somebody tries to spoil the survey with deceitful responses, the survey should be so designed that it may not be ruined. For instance, 10–15% extra respondents may be included in the survey, so that such cases can be eliminated.

3.2.6 Prior Knowledge of the Respondents

It is assumed that the respondents do not have the prior knowledge of acronyms and jargon used in the survey. Thus, if such words and phrases are used without explaining their meaning, the responses are bound to be misleading. We have seen many of the surveys in which the researchers use lots of acronyms and jargons under the pretext that the respondents have prior knowledge about these wordings. In such cases, responses of the survey may be skewed and inaccurate.

3.2.7 Clarity About Items in the Questionnaire

It is assumed that the respondents understand each question in the survey correctly. Otherwise, the results may be misleading and will have an adverse effect on the findings. For example, if you ask a person for his or her nationality, it may not be clear as to what you want. Do you want someone from India to say Indian, Hindu, or Aryan? Similarly, if you ask about marital status, do you wish the respondent to say simply married or unmarried? Or do you wish them to give more detail like divorced, widow/widower, etc.? Since each questionnaire is meant for a target population, a pilot run and feedback in the survey study is advisable to ensure that the respondents understand each question in the questionnaire correctly.

3.2.8 Ensuring Survey Feedback

It is assumed that the respondents will receive the feedback of the survey or at least they should be able to see it in the publication. Assurance of feedback not only ensures seriousness observed by the participants but also enhances reliability of information obtained in the survey. The researcher must publish the findings or upload the report on the website if possible and inform all the participants in the survey about it. This encourages the respondents to not only take part voluntarily in future surveys organized by the researcher but also provide the correct information.

3.2.9 Nonresponse Error

It is assumed in the survey studies that there is least nonresponse error. This is one of the major causes of misleading findings in the survey research.

The magnitude of nonresponse errors becomes sizable, especially in electronic and postal survey. Due to nonresponse errors, a particular segment of the population's representation may not be there in the survey, resulting in skewed findings. Various measures should be taken to minimize the nonresponse errors for correct findings.

3.3 Questionnaire's Reliability

One of the most important assumptions in survey research is that the questionnaire is reliable. Generally, reliability means dependability, but in the research context, it refers to the repeatability or consistency. As per Anastasi (1982), reliability can be defined as "consistency of scores obtained by the same individuals when reexamined with test on different occasions, or with different sets of equivalent items, or under other variable examining conditions."

A questionnaire is considered to be reliable if it gives the same response in repetitive testing (assuming testing conditions remain same). Repeatability or consistency of questionnaire can be assessed by the test–retest methods. This method of test–retest actually measures temporal reliability of the questionnaire. If a questionnaire is designed to measure state anxiety, then it must give similar results if tested repeatedly on the same subject. In other words, if a questionnaire is administered, all respondents retain their relative ranks of two separate measurements with the same test.

Further reliability also means internal consistency of a measure in the questionnaire. Respondents who obtained high on one set of items, also score high on an equivalent set of items and vice versa. Internal consistency can be measured by the split-half test. The internal consistency of the questionnaire can be estimated with greater accuracy by any of the two methods, namely Kuder–Richardson or Cronbach's alpha, depending upon the response type in the questionnaire.

We shall now discuss the procedure of estimating two different types of reliability, i.e. repeatability and internal consistency.

3.3.1 Temporal Stability

Temporal stability refers to the repeatability of responses in the questionnaire. Temporal stability of the questionnaire ensures consistency between the responses obtained of the same persons if tested over a period of time. The correlation coefficient between the responses indicates temporal stability and is known as coefficient of stability. The coefficient of stability is estimated by the test–retest method.

3.3.1.1 Test–Retest Method

This test measures consistency of assessment over a period of time and is a measure of temporal stability. In this test, responses are obtained on same subjects at two different times. This kind of reliability is used to assess the consistency of a test across time. In applying this test, we assume that there will be no substantial change in the construct being measured between the two occasions. Administering the same measure on two different times provides two sets of scores. The coefficient of correlation obtained on these two sets of scores is the reliability coefficient. This reliability coefficient is also known as *temporal stability coefficient*. It reveals as to what extent the respondents retain their relative positions as measured in terms of test score over a given period of time. If the respondents who obtain low (high) scores on the first administration also obtain low (high) scores on the second administration, the correlation coefficient between the two sets of score (test and retest) will be high. Higher the value of correlation, more reliable the test is. In test–retest method, the time interval between the first and second testing is a critical issue. However, two weeks of gap is considered to be most appropriate. Reliability of the questionnaire estimated by the test–retest method is inversely proportional to the time gap between the two testing.

3.3.2 Internal Consistency

Internal consistency refers to the consistency of responses obtained in a test, such that the different items measuring the construct provide consistent scores. According to Freeman (1953), reliability of a test is its ability to yield consistent results from one set of measures to other. If a lifestyle assessment test consisting 30 questions is divided into two groups, then the test would be said to have internal consistency if the correlation between the two sets of responses is high and positive. This will happen only when both the group of items measures the same construct, i.e. lifestyle. It is difficult to test reliability by means of internal consistency because exact measurements of the psychological parameters are not possible. However, it can be estimated in a number of ways by the tests such as split-half, Kuder–Richardson, and Cronbach's alpha. We shall explain these tests in the following sections.

3.3.2.1 Split-Half Test

This is the most popular method used for measuring internal consistency. Split-half test is a measure of internal consistency, which reveals how well the test components contribute to the construct that is measured. In this test, items of the questionnaire are divided randomly (preferably in even and odd questions) into two sections and then both sections of the test are given to one group of subjects. The scores from both the sections of the test are correlated. A reliable questionnaire will have high correlation, indicating that a subject would perform equally well (or as poorly) on both sections of the questionnaire. The main advantage of the split-half method is that there is no need to administer the test twice because all the test items are divided into two groups.

A single administration of test yields all data needed for the computation of reliability coefficient. This test is most commonly used for multiple-choice tests. The drawback of the split-half test is that it cannot be used with a speed test. Another disadvantage of this method is that if the test is divided into two halves by some other method instead of dividing into even and odd test items, reliability coefficient differs. In other words, different methods of dividing the test items yield different reliability coefficients. Another drawback of the test is that it only works for large set of questions, say 100 or more. Following are the steps in applying this test.

Steps

1) Randomly divide the items in the questionnaire into two sections.
2) Ask a group of subjects to respond to the items in both the sections in a single sitting.
3) Determine the score for each subject in each half of the test.
4) Compute correlation coefficient for the data in the two sections.

Illustration 3.1 Let us see the procedure involved in test–retest method. Consider a questionnaire measuring lifestyle of the subject in a specific population. Each statement has three response options and the respondent can choose any one of them. The test consists of six statements only as shown in Table 3.1. In order to apply the split-half test, these six statements have been classified into two groups. In the first group, statements 1, 3, and 5 are kept, whereas in the second group, statements 2, 4, and 6 are included. Table 3.1 shows the response of a typical subject on both the sets of statements. It can be seen that the subject score in the first half is 9 (statements 1, 3, and 5) and in the second half 7 (statements 2, 4, and 6). Thus, this subject's score in each

Table 3.1 A typical scoring of the subject's response in a lifestyle assessment test.

Statements		<2 days	2–4 days	>4 days
1.	How often you take fried food in a week?	3		
2.	How many days in a week you take sprouts in your food?		2	
3.	How many days in a week you work out 30 minutes a day?			3
4.	How many days you take alcohol in a week?		2	
5.	How often you sleep more than six hours in a day?			3
6.	How many days you smoke in a week?	3		

Table 3.2 Scores of the subjects in each group of statements.

Subject	Odd group (1)	Even group (2)
1	9	7
2	6	5
3	7	6
4	4	3
5	6	5
6	8	6
7	6	5
8	7	6
9	6	5
10	7	8
	$r = 0.817$	

of the two sets of statements are 9 and 7. Table 3.2 indicates the response of 10 subjects on the two sets of statements. The correlation coefficient for this data set is 0.817. This is the reliability coefficient of the questionnaire measuring lifestyle. In fact, the magnitude of the correlation tells us about the extent of internal consistency of the questionnaire.

3.3.2.2 Kuder–Richardson Test

Kuder–Richardson test is a measure of reliability for a test in which responses of the questions are binary in nature. This test can only be used when there is a correct answer for each question and should not be used if the responses are based on Likert scale. Reliability of a test ensures that the test actually measures what it supposes to measure. There are two versions of the Kuder–Richardson test: KR-20 and KR-21. If items in the test have varying difficulty, then KR-20 formula is used, and if all the questions are equally challenging, then one should use KR-21 formula. These formulas are as follows:

$$\text{KR-20} = \left(\frac{n}{n-1}\right) \times \left(1 - \frac{\sum(p \times q)}{V}\right) \tag{3.1}$$

$$\text{KR-21} = \left(\frac{n}{n-1}\right) \times \left(1 - \frac{M \times (n-M)}{n \times V}\right) \tag{3.2}$$

where n is the sample size, p is the proportion of the subjects passing the item, q is the proportion of the subjects failing the item, and M and V are the mean and variance of the test, respectively.

Table 3.3 English test with possible answers.

1.	She was unjustified in accusing us ….. theft.	i. for	ii. of
2.	He tries to adjust ….. his relations.	i. with	ii. at
3.	We all are devoted ….. one another.	i. to	ii. for
4.	Work hard ….. you should fail.	i. otherwise	ii. lest
5.	I worked him ….. into a great passion.	i. up	ii. upon
6.	He persisted ….. doing the job despite its being uninteresting to him.	i. at	ii. in
7.	She had left him ….. the lurch.	i. on	ii. in
8.	He congratulated her …. her brilliant success.	i. upon	ii. on
9.	She made my position clear ….. her face.	i. on	ii. to
10.	He jumped ….. the river to have a dip.	i. in	ii. into
11.	While I was speaking, he kept cutting ……	i. in	ii. down
12.	She said that she ….. attend the meeting.	i. would	ii. shall
13.	When I was at Delhi, I ….. exercise everyday.	i. took	ii. would take
14.	Edwin is the ….. boy in his class.	i. good	ii. best
15.	If she ….. helped you, you would have passed.	i. has	ii. had

Kuder–Richardson coefficient ranges from 0 to 1, indicating 0 as no reliability and 1 as perfect reliability. As the coefficient increases toward 1, the reliability of the test also increases. A test can be considered to be reliable if the value of the coefficient is 0.5 or more.

Illustration 3.2   A test consisting of 15 questions was developed to assess the English knowledge of the high school boys. Items in the test have varying difficulty. Each item in the test has one correct answer. This test was administered on 30 subjects, and they were asked to complete the test in 15 minutes. All 15 questions along with 2 possible answers are shown in Table 3.3. We shall check the reliability of this test using the Kuder–Richardson formula.

Since items in the test have varying difficulty, we shall use KR-20 formula to compute the reliability coefficient.

Steps

Computation of reliability coefficient has been shown in Table 3.4. By following the below-mentioned steps, we can compute KR-20 coefficient.

1) In the first column, write the number of subjects who have responded the item correctly.
2) In the second column, write the number of subjects who have responded the item wrongly.

Table 3.4 Computation in Kuder–Richardson test for testing reliability.

Item	Description	No. of correct responses (1)	No. of wrong responses (2)	Proportion of correct responses (p) (3)	Proportion of wrong responses (q) (4)	p*q (5)
1	She was unjustified in accusing us ….. theft.	13	17	0.43	0.57	0.25
2	He tries to adjust …. his relations.	5	25	0.17	0.83	0.14
3	We all are devoted ….. one another.	12	18	0.4	0.6	0.24
4	Work hard ….. you should fail.	7	23	0.23	0.77	0.18
5	I worked him ….. into a great passion.	12	18	0.4	0.6	0.24
6	He persisted ….. doing the job despite its being uninteresting to him.	6	24	0.2	0.8	0.16
7	She had left him ….. the lurch.	14	16	0.47	0.53	0.25
8	He congratulated her …. her brilliant success	6	24	0.2	0.8	0.16
9	She made my position clear …. her face.	5	25	0.17	0.83	0.14
10	He jumped ….. the river to have a dip.	12	18	0.4	0.6	0.24
11	While I was speaking, he kept cutting ……	13	17	0.43	0.57	0.25
12	She said that she …. attend the meeting.	12	18	0.4	0.6	0.24
13	When I was at Delhi, I ….. exercise everyday.	15	15	0.5	0.5	0.25
14	Edwin is the …. boy in his class	11	19	0.37	0.63	0.23
15	If she ….. helped you, you would have passed.	9	21	0.3	0.7	0.21
	$n = 30$	variance =11.18				$\sum p \times q = 3.17$

3) In the third column, write the proportion of the subjects who have responded the item correctly.
4) In the fourth column, write the proportion of the subjects who have responded the item wrongly.
5) In the fifth column, multiply p and q for each item.
6) Find the variance of the scores in column 1, which is 11.18 in this illustration.
7) Find the total of the fifth column to get $\sum p \times q$.
8) Substitute the values in formula (3.1) to get the reliability coefficient, which is 0.768.

$$\text{KR-20} = \left(\frac{n}{n-1}\right) \times \left(1 - \frac{\sum p \times q}{\text{Var}}\right)$$

$$= \left(\frac{30}{30-1}\right) \times \left(1 - \frac{3.17}{11.18}\right) = 0.768$$

3.3.2.3 Cronbach's Alpha

Cronbach's alpha is a measure of reliability or internal consistency that a questionnaire has. The meaning of reliability is how well a questionnaire measures what it should measure. For instance, if a questionnaire is meant for testing motivation of the employee, then high reliability means it measures motivation, while low reliability indicates that it measures something else. The Cronbach's alpha was developed by Cronbach (1951) and is also denoted by α. This test is used for testing reliability of questionnaires whose responses are on Likert scale. The responses may have three or more options. In fact, questionnaires are developed to measure the latent variables that are difficult to measure directly; hence, different types of statements are kept in them. A researcher may develop the questionnaire for measuring lifestyle, personality, frustration, neurosis, etc. The Cronbach's alpha indicates whether the set of statements you have selected in the questionnaire are reliable and actually measures what it is meant to measure. While calculating Cronbach's alpha, we assume that the questionnaire is measuring only one construct or dimension. If more than one dimension is measured in questionnaire, then the test results would be meaningless. In that case, the questions can be categorized into parts, measuring different constructs. In case it is difficult to categorize, then factor analysis may be used to identify dimensions in the questionnaire.

The Cronbach's alpha (α) is computed by the following formula:

$$\alpha = \frac{n\bar{c}}{\bar{v} + (n-1)\bar{c}} \tag{3.3}$$

where n is the number of items in the questionnaire, $\bar{c}$ is the average covariance between the item pairs, and $\bar{v}$ is the average variance.

Computing Cronbach's alpha using the formula may be a cumbersome process for the researcher; hence, we shall discuss its computation with IBM SPSS® Statistics software ("SPSS")[1] by means of Illustration 3.3.

Illustration 3.3 Consider a questionnaire measuring lifestyle of an individual having 10 statements as shown in Table 3.5. Each question can be answered on 5-point Likert scale with range 1 (least suitable) to 5 (most suitable) for good lifestyle. We shall see the procedure for computing Cronbach's alpha with SPSS for this questionnaire.

Let us suppose that the questionnaire has been administered on 17 subjects randomly selected from the population of interest. The score obtained by these subjects are shown in Table 3.6.

After preparing the data file in SPSS, follow the sequence of commands **Analyze → Scale → Reliability Analysis** as shown in Figure 3.1. The procedure of making the data file has been discussed in Chapter 2.

After pressing **Reliability Analysis** command, Figure 3.2 shall be obtained to select the variables and the options for the analysis. Shift all the 10 variables from left panel to the "Items" section in the right panel of the screen. Click on **Statistics** to get Figure 3.3 and then check "Item," "Scale," and "Scale if Item deleted." Let other options remain selected by default. Click on **Continue** and **OK** to get the output shown in Tables 3.7 and 3.8.

Table 3.7 indicates the value of Cronbach's alpha as 0.822. We shall explore whether this value can be enhanced. Let us look at the second output in

Table 3.5 Sample questions on a lifestyle assessment test.

Item	Description
1.	How many days in a week you socialize with your friends?
2.	How many days in a week you take sprouts in your food?
3.	How many days in a month you visit to the religious organizations?
4.	How many days you smoke in a week?
5.	How many days you take alcohol in a week?
6.	How many days in a week you sleep more than six hours a day?
7.	How often you take fried food in a week?
8.	How many days in a week you work out 30 minutes a day?
9.	How many times in a year you were sick?
10.	How many days in a week you prepare yourself before sleep?

---

1 SPSS Inc. was acquired by IBM in October, 2009.

Table 3.6 Response of the subjects on different items of the questionnaire.

Subject	Items									
	I1	I2	I3	I4	I5	I6	I7	I8	I9	I10
1	2	3	2	3	1	2	3	3	2	1
2	3	1	4	1	1	4	3	2	1	1
3	4	2	3	3	4	1	2	4	3	4
4	1	3	2	1	3	2	4	3	3	3
5	4	4	2	3	5	4	5	5	4	4
6	3	4	4	2	1	2	3	2	1	1
7	2	4	3	4	3	4	3	2	3	3
8	4	3	3	4	3	4	3	3	3	3
9	4	4	4	4	5	5	5	3	4	4
10	2	4	4	2	4	4	4	4	4	4
11	1	2	3	4	2	3	4	3	4	2
12	4	3	4	3	4	4	3	3	4	4
13	2	4	3	4	4	5	3	4	3	4
14	2	3	4	1	3	4	4	2	4	3
15	3	4	2	4	2	3	3	1	4	2
16	2	4	4	3	2	4	4	4	4	4
17	5	5	4	3	4	4	5	4	4	4

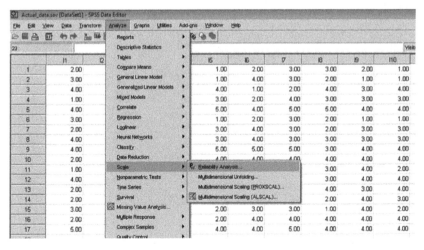

Figure 3.1 Screen showing commands for reliability analysis. Source: Reprint Courtesy of International Business Machines Corporation, © International Business Machines Corporation.

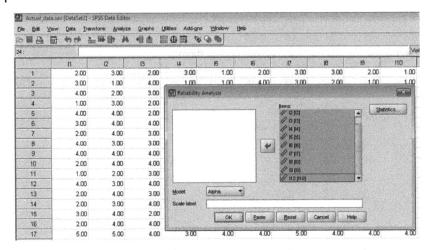

Figure 3.2 Screen showing selection of items for reliability analysis. Source: Reprint Courtesy of International Business Machines Corporation, © International Business Machines Corporation.

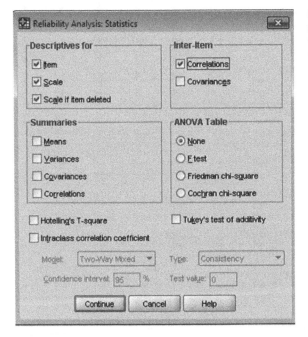

Figure 3.3 Screen showing options for reliability outputs. Source: Reprint Courtesy of International Business Machines Corporation, © International Business Machines Corporation.

Table 3.7 Reliability statistics.

Cronbach's alpha	Cronbach's alpha-based on standardized items	No. of items
0.822	0.814	10

Table 3.8 Output of the reliability analysis.

Items		Scale means if item deleted	Scale variance if item deleted	Corrected item total correlation	Squared multiple correlation	Cronbach's alpha if item deleted
1	How many days in a week you socialize with your friends?	28.82	37.90	0.36	0.41	0.823
2	How many days in a week you take sprouts in your food?	28.29	37.35	0.51	0.51	0.806
3	How many days in a month you visit to the religious organizations?	28.41	42.26	0.14	0.53	0.835
4	How many days you smoke in a week?	28.76	39.44	0.27	0.56	0.829
5	How many days you take alcohol in a week?	28.65	31.12	0.79	0.88	0.769
6	How many days in a week you sleep more than six hours a day?	28.18	36.15	0.52	0.57	0.804
7	How often you take fried food in a week?	28.06	38.31	0.50	0.78	0.807
8	How many days in a week you work out 30 minutes a day?	28.59	37.26	0.49	0.74	0.807
9	How many times in a year you were sick?	28.41	35.51	0.64	0.82	0.791
10	How many days in a week you prepare yourself before sleep?	28.65	32.24	0.82	0.95	0.768

Table 3.9 Guidelines for internal consistency benchmark.

Cronbach's alpha (α)	Internal consistency
$\alpha < 0.5$	Unacceptable
$0.5 \leq \alpha < 0.6$	Poor
$0.6 \leq \alpha < 0.7$	Questionable
$0.7 \leq \alpha < 0.8$	Acceptable
$0.8 \leq \alpha < 0.9$	Good
$\alpha \geq 0.9$	Excellent

Table 3.8 generated in SPSS analysis. In the last column, you can notice that if item number 3 is deleted, then Cronbach's alpha can be enhanced to 0.835. This is an indication that item number 3 is not consistent with the other items in the questionnaire to measure the lifestyle of an individual. Similarly, it can also be noticed that item numbers 4 and 1 will also enhance the value of α from its current level if they are eliminated. Since this type of analysis is based on empirical data only, one should also look into the theoretical aspects in measuring the construct. For instance, if theory suggests that the socialization is good for the healthier lifestyle, then item number 1 should not be deleted even if it increases the α by its removal.

Cronbach's alpha ranges from 0 to 1, with higher values indicating greater internal consistency. In general, any value of α greater than 0.7 may be considered acceptable for the questionnaire to be reliable. As a rule of thumb, one can consider the guidelines for internal consistency as shown in Table 3.9.

Remark The questionnaire discussed in 3.3 is not the actual one; hence, no conclusion should be drawn on the basis of the findings discussed in this illustration. It has been used merely for the purpose of explaining the procedure of computing α with SPSS.

Exercise

Multiple-Choice Questions

Note: Choose the most appropriate answer.
1. In survey studies, sample size depends upon
   a. Accuracy of estimate
   b. Variability of the population characteristics
   c. Confidence coefficient
   d. All of the above

2. Reason for defining delimitation in research studies is
   a. To have the ease in conducting study
   b. To control the variability of population for reliable findings
   c. To reduce the time in the study
   d. To reduce the cost in the study

3. If assumptions in survey studies are not met, then
   a. Estimate may be overestimated or underestimated
   b. Increasing sample size may compensate the accuracy
   c. Carefully testing with small sample will not affect the findings
   d. Confidence interval of estimating parameter will decrease

4. Data cleaning in research studies is not done for
   a. Identifying the outliers
   b. Consistency check
   c. Treating missing data
   d. Removing wrongly entered data

5. Nonresponse error in survey studies arises due to
   a. Respondents not cooperating with the researcher
   b. Respondents giving wrong information
   c. Respondents not receiving the questionnaire
   d. Respondents not participating after identifying

6. Reliability of the questionnaire refers to
   a. Dependability of questionnaire
   b. Repeatability of questionnaire
   c. Using it for any population
   d. Providing correct findings even with the small samples

7. Test–retest method measures
   a. Temporal reliability of questionnaire
   b. Internal consistency of questionnaire
   c. Objectivity of questionnaire
   d. Validity of questionnaire

8. Which of the statement is not correct?
   a. Test–retest method is used for multiple choice tests.
   b. Split-half test cannot be used with a speed test.
   c. Split-half test results in different coefficients depending upon how the statements are split.
   d. Test–retest method can be used with small set of questions.

9. Kuder–Richardson test is used when the
    a. Responses of the questions are measured on ordinal scale.
    b. Responses of the questions are on 3-point Likert scale.
    c. Responses of the questions are either true or false.
    d. Responses of the questions are on 5-point Likert scale.

10. Which of the following statement is true in relation to reliability test?
    a. KR-20 is used if questions have varying difficulty.
    b. KR-20 formula is used if all the questions are equally challenging.
    c. KR-21 formula is used if questions have varying difficulty.
    d. KR-21 is used when responses are on the Likert scale.

11. Choose the correct formula for Kuder–Richardson test (KR-20).
    a. $\left(\dfrac{n}{n+1}\right) \times \left(\dfrac{\sum(p \times q)}{V}\right)$
    b. $\left(\dfrac{n-1}{n}\right) \times \left(1 - \dfrac{\sum(p \times q)}{V}\right)$
    c. $\left(\dfrac{n}{n-1}\right) \times \left(\dfrac{\sum(p \times q)}{V}\right)$
    d. $\left(\dfrac{n}{n-1}\right) \times \left(1 - \dfrac{\sum(p \times q)}{V}\right)$

12. A test may be considered to be reliable if the Kuder–Richardson coefficient is
    a. 0
    b. 0.5 or more
    c. 0.4
    d. More than 0.3

13. Cronbach's alpha can be used to test the reliability of the questionnaire only if it is meant for testing
    a. One construct
    b. Two constructs
    c. Three constructs
    d. Any number of constructs

14. A questionnaire may be considered to be reliable only if Cronbach's alpha is
    a. More than 0.5
    b. More than 0.7
    c. 0.6
    d. 0.4

15. Which is the correct formula for KR-21?

a. $\left(\dfrac{n-1}{n}\right) \times \left(1 - \dfrac{M \times (n-M)}{n \times V}\right)$

b. $\left(\dfrac{n}{n-1}\right) \times \left(1 - \dfrac{M \times (n-M)}{n \times V}\right)$

c. $\left(\dfrac{n}{n-1}\right) \times \left(\dfrac{M \times (n-M)}{n \times V} - 1\right)$

d. $\left(\dfrac{n}{n-1}\right) \times \left(1 + \dfrac{M \times (n-M)}{n \times V}\right)$

Short-Answer Questions

1. Whether sample should always be preferred over population in conducting survey research even if there are no resource considerations? Explain your views.

2. What do you understand by nonresponse error? Does it affect the hypothesis testing experiments?

3. What are the important assumptions regarding respondents while conducting survey?

4. What do you mean by reliability of the questionnaire? Differentiate between test–retest method and split-half method of estimating reliability.

5. Explain the situations in which the Kuder–Richardson test and Cronbach's alpha are applied. What is the difference between these two methods of estimating reliability?

Answers

Multiple-Choice Questions

1. d

2. b

3. a

4. c

5. d

6. b

7. a

8. d

9. c

10. a

11. d

12. b

13. a

14. b

15. b

4

Assumptions in Parametric Tests

4.1 Introduction

"Statistics is a branch of mathematics dealing with the collection, analysis, interpretation, presentation, and organization of data" (Triola 2001). Another definition of statistics is the art and science of learning from the data. Statistical analyses can be divided into two categories: descriptive statistics and inferential statistics. Descriptive statistics comprises the methods for data representation by calculating descriptive figures, creating the summary graphs, or both. Descriptive statistics commonly uses figures/tables that include the frequencies, percentages, measures of central tendency, and measures of variation. Each of these descriptive figures uniquely summarizes the data (by various estimators) for better understanding the characteristics of data in hand. Also, the descriptive analysis gives useful information before starting the statistical inference, subject to the assumption that the sample is randomly drawn from the target population.

The results of descriptive analysis help the researchers and practitioners to have an overview and some reasonable idea about the population from which the random sample is drawn. The sample descriptions are then used to infer some information about population with some uncertainty by testing the stated research hypothesis and constructing the confidence interval for the respective parameters based on their estimates. The statistical procedures can thus be regarded as scientific methods, rather than just guessing about the population. However, these procedures cannot be applied without making some assumptions regarding the probability distribution of the random variable used in the inferential procedure. This makes some of the estimators performing better than others in certain situations, based on the underlying population distribution.

The population properties are therefore important to be accessed in choosing the best statistical model/method for the data under consideration based on the set of stated assumptions. These set of assumptions make a distinction

Testing Statistical Assumptions in Research, First Edition. J. P. Verma and Abdel-Salam G. Abdel-Salam.
© 2019 John Wiley & Sons, Inc. Published 2019 by John Wiley & Sons, Inc.
Companion Website: www.wiley.com/go/Verma/Testing_Statistical_Assumptions_Research

between two different procedures for inferential statistics, called parametric and nonparametric techniques. Usually, the parametric tests are known to be associated with strict assumptions about the underlying population distribution. On the other hand, the nonparametric tests are considered as distribution-free inferential methods, where the data does not comply with the common assumption for normal distribution. However, it does not make the nonparametric tests to be assumption free but less conservative.

The choice between parametric and nonparametric tests does not solely depend on the assumption of normality; this assumption can be relaxed for large sample size, backed up by the central limit theorem. Rather, the choice also depends on the level of measurement of the variables under consideration.

The parametric tests are usually meant for the variables measured on interval/ratio scale, while nonparametric tests are applicable for nominal and ordinal data. According to Sheskin (2003), when the data is measured over interval or ratio scale, the parametric tests should be tried first. However, if there is any violation for one or more of the parametric assumptions, it is recommended to transform the data into a format that makes it compatible to the appropriate nonparametric test.

4.2 Common Assumptions in Parametric Tests

Every statistical test (whether parametric or nonparametric) has several assumptions about the data on which the test is intended to be used. The parametric tests usually require more assumptions than the nonparametric tests and are therefore considered as more conservative. The unmet assumptions may lead the researchers/practitioners to draw wrong conclusions about the results, and the inferences implied will not be justified. However, some of the parametric methods are "robust" to certain assumptions while keeping the results reliable. Some of the common assumptions for the parametric tests include Normality, Randomness, Absence of Outliers, Homogeneity of Variances, Independence of Observations, and Linearity. In the following sections, details about each assumption and how to perform using IBM SPSS[1] Statistics software (SPSS) shall be discussed.

4.2.1 Normality

For almost all of the parametric tests, a normal distribution is assumed for the variable of interest in the data under consideration. Normality is one of the essential assumptions for drawing reliable inferences about the underlying population of data. However, this assumption could be relaxed for large

---

1 SPSS Inc. was acquired by IBM in October 2009.

sample size, where the researcher believes that large sample represents similar characteristics as of the population. Normal distribution is a symmetric bell-shaped curve, showing that the observation in the data is symmetrically distributed and there is a minimum chance for bias. This type of data is symmetric around its mean value and has a kurtosis equivalent to zero. When testing for normality, we provide tests to determine whether the data meets the assumption.

The perfect normality of data arrives when skewness is "zero," which is practically rare. The variable of interest deviating from perfect normality can be said as approximate normal and the parametric tests are applicable within the range of ±1 skewness. Another measure of normality is "kurtosis," which depicts the normal density of data by the peakedness of normality curve. In the following section, the readers will get familiar with steps in testing the normality utilizing the SPSS software.

4.2.1.1 Testing Normality with SPSS

The normality assumption can be tested using the SPSS by specific statistical tests and graphical representation of the variable of interest. The common statistical tests for normality in the univariate case are the Kolmogorov–Smirnov test, which is preferable for large samples, and Shapiro–Wilk test, for small to medium samples. The null hypothesis for normality testing is stated as "the data is normal."

H_0: The distribution is normal.
H_a: The distribution is NOT normal.

The graphical methods include histogram, stem-and-leaf diagram, normal Q–Q plot, P–P plot, and box plot. The procedure for testing normality with SPSS has been explained by using the dataset in Table 4.1.

The normality assumption can be tested using the Shapiro–Wilk test or Kolmogorov–Smirnov tests in SPSS. Shapiro–Wilk test is more suitable for small samples ($n \leq 50$) but it can be used for the sample sizes up to 2000 observations, whereas Kolmogorov–Smirnov test is used for large samples. One of the weaknesses of these tests is giving significant results for the large sample even for slight deviation from normality.

We shall show the procedure of testing normality of data with SPSS by means of an example. Let us consider the scores obtained by the students out of 10 in three different subjects as shown in Table 4.1.

Let us see how normality can be tested using the SPSS software. To start with, a data file needs to be prepared in SPSS using the data shown in Table 4.1. The procedure of making a data file has been discussed in Chapter 2. After preparing the data file, follow the sequence of commands, **Analyze → Descriptive Statistics → Explore**, as shown in Figure 4.1.

Table 4.1 Performance of the students.

Maths	English	Science
8.0	8.0	5.0
3.0	7.0	7.0
4.0	6.0	7.0
3.0	4.0	2.0
3.0	6.0	6.0
4.0	3.0	5.0
5.0	5.0	8.0
5.0	6.0	6.0
4.0	4.0	1.0
4.0	7.0	7.0
3.0	5.0	8.0
3.0	2.0	6.0

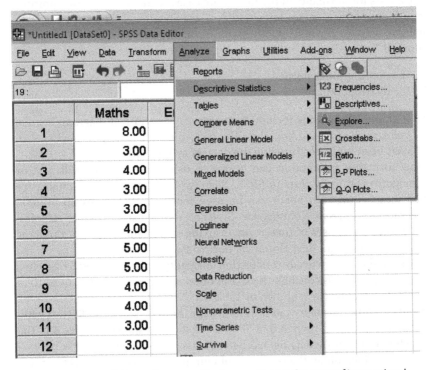

Figure 4.1 Commands for testing normality. Source: Reprint Courtesy of International Business Machines Corporation, © International Business Machines Corporation.

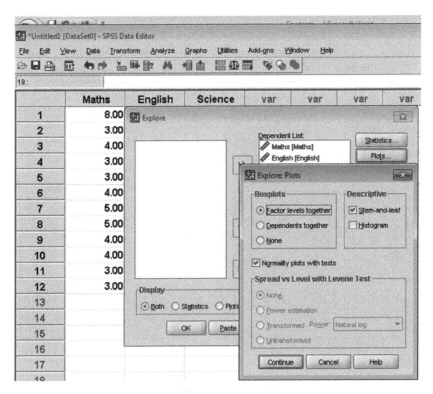

Figure 4.2 Screen showing options for computing Shapiro–Wilk test. Source: Reprint Courtesy of International Business Machines Corporation, © International Business Machines Corporation.

After pressing **Explore** command, the screen as shown in Figure 4.2 shall be obtained to select the variables and the options for the analysis. Shift all the three variables from the left panel to the "Dependent List" section of the screen. Click on **Plots** and check the "Normality plots with test." This option will give you the outputs of Kolmogorov–Smirnov and Shapiro–Wilk tests along with the Q–Q plot. The Q–Q plot is a graphical way of showing normality condition of the data. Let other options remain selected by default. Click on **Continue** and **OK** to get the output shown in Table 4.2 and Figure 4.3.

Table 4.2 shows the Kolmogorov–Smirnov and Shapiro–Wilk test statistics for testing normality of data. Normality exits if these tests are not significant. Thus, if the significance value (p-value) of these tests is more than 0.05, the data is considered to be normal; otherwise, normality assumption is violated. Looking at the values of these tests in Table 4.2, it may be concluded that the scores on Maths ($p < 0.05$) and Science ($p < 0.05$) violates normality, whereas normality exists for the scores on English ($p > 0.05$). This is so because the Shapiro–Wilk test statistic for the scores on Maths and Science is significant, whereas for English, it is nonsignificant.

Table 4.2 Tests of normality for the data on students' performance.

	Kolmogorov–Smirnov			Shapiro–Wilk		
	Statistics	df	Sig. (*p*-value)	Statistic	df	Sig. (*p*-value)
Maths	0.273	12	0.014	0.742	12	0.002[a)]
English	0.165	12	0.200	0.968	12	0.893
Science	0.227	12	0.088	0.854	12	0.041[a)]

a) Significant at 5% level.

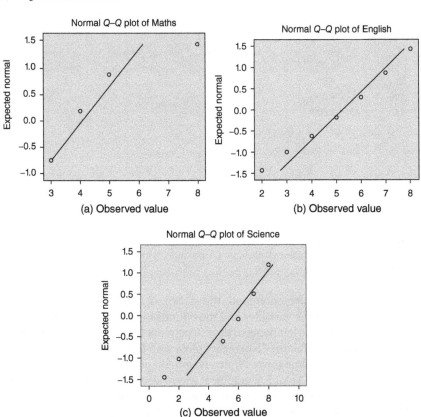

Figure 4.3 Normal we Q–Q plot for the data on student's performance.

Q–Q Plot for Normality Q–Q plot is a graphical way of showing normality of the data by comparing two probability distributions: the observed data and a generated standard normal data by plotting their quintiles against each other. If the distribution of the observed sample is similar to that of the standard normal distribution, the points in the Q–Q plot will lie on the line. The line indicates that the value of your sample should fall on it or close by if the data follows normal distribution. Deviation of these points from the line indicates non-normality. The normal Q–Q plot obtained on the scores mentioned in Table 4.1 is shown in Figure 4.3. In Figure 4.3a,c, the data deviates from the line indicating that the scores on Maths and Science violate normality, whereas in Figure 4.3b, the data fall along the line, more closely indicating normality of the scores on English.

4.2.1.2 What if the Normality Assumption Is Violated?

We have seen earlier that for metric data, the parametric statistics and for nonmetric data, the nonparametric statistics are computed. But parametric statistics provides correct picture only when the metric data is normally distributed. Let us see what happens if we compute the parametric statistics when the normality assumption is violated. Consider the weight in kilograms of the 14 students as shown below:

| X | 51 | 52 | 55 | 59 | 60 | 61 | 61 | 62 | 62 | 63 | 72 | 71 | 82 | 90 |

Since this is a scale data, we can compute parametric statistic for central tendency using the mean. The mean of this data set is 64.36. It can be seen that most of the scores are less than 64.36; hence, it cannot be considered as the true representative of the data set. Why it is so, let us examine the data set by plotting the histogram as shown in Figure 4.4. It can be seen that the data is positively skewed and lacks symmetricity; hence, it violates normality. By applying Shapiro test, it can be seen that normality is violated.

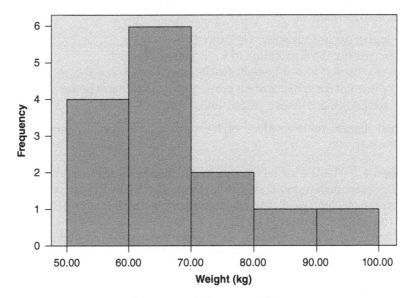

Figure 4.4 Distribution of data on weight by means of histogram.

Since normality assumption has been violated, we need to compute nonparametric statistic for central tendency. Let us now compute the nonparametric statistic, median for the data on weight. Here, there are 14 observations ($n = 14$); hence,

$$\text{Median} = \left(\frac{n+1}{2}\right)^{\text{th}} \text{score} = 61.5.$$

It can be seen that 61.5 as an average is a better indicator of average than 64.36. Since most of the hypothesis testing experiment is done on mean and if the mean does not indicate the correct picture of the scores due to violation of normality, then statistical tests may give wrong conclusions. So, the researchers and practitioners should try some transformations on the data to make it normal if possible, otherwise use the nonparametric statistics. Here are some commonly used transformations for converting non-normal data into normal.

4.2.1.3 Using Transformations for Normality

The parametric statistical methods assume that the variable of interest (dependent variable) should be normally distributed. The non-normal data can be transformed to attain normality using the box-cox transformation or power transformation, where the central limit theorem cannot justify the small sample size ($n < 30$ observations for each estimated parameter). The outliers however should be deleted first (one at a time) if these outliers are not giving any extra information and are just randomly appearing in the data. Following transformations suitable in different situations can be used to convert the skewed data to the normal distribution:

- $f(x) = \log(x)$; use it if x is positively skewed.
- $f(x) = x^2$; use it if the distribution of x is negatively skewed.
- $f(x) = \sqrt{x}$; use it if x has a Poisson distribution ($x = 0, 1, 2, \ldots$).
- $f(x) = \frac{1}{x}$; use it if the variance of x is proportional to the fourth power of $E(x)$.
- $f(x) = \arcsin(x)$; use it if x is a proportion or rate.

We shall discuss the procedure of transforming the data by means of Illustration 4.1.

Illustration 4.1   Table 4.3 shows the marks obtained by the students in Maths and English examinations in a class. Let us see whether these data sets are normally distributed. If not, we shall try some appropriate transformations to make them normal.

Let us first check whether the data sets X and Y are normally distributed. By adopting the procedure of testing normality with SPSS software as discussed in Section 4.2.1, the outputs can be obtained, which is shown in Table 4.4. This table reveals that Shapiro statistics for X ($p < 0.05$) as well as Y ($p < 0.05$) are significance at 5% level.

Table 4.3 Marks of the students.

Maths (X)	English (Y)
2	6
3	8
3	12
4	18
2	14
3	15
5	16
4	13
6	18
7	16
4	17
8	15
12	4
14	17

Table 4.4 Tests of normality for the data on students' marks.

	Kolmogorov–Smirnov			Shapiro–Wilk		
	Statistics	df	Sig. (p-value)	Statistic	df	Sig. (p-value)
Maths (X)	0.231	14	0.042	0.832	14	0.013[a]
Science (Y)	0.202	14	0.125	0.856	14	0.027[a]

a) Significant at 5% level.

Now we shall apply the relevant transformations to X and Y in order to check whether the transformed variable will have normal distribution. In order to do so, first we shall check the distribution of these two variables by means of the histogram. Figure 4.5 shows that X is positively skewed; hence, as per the guidelines mentioned above, we require log transformation to make it normal. Similarly, the distribution of Y in Figure 4.6 seems to be negatively distributed; it requires square transformation. Thus, converting X into $\log(X)$ and Y into Y^2, we can transform these two variables. The variables and their transformed scores are shown in Table 4.5.

In order to check whether the distribution of these transformed variables $\log(X)$ and Y^2 becomes normal, we shall again apply Shapiro–Wilk test, the output of which is shown in Table 4.6. It can be seen that the Shapiro statistics for X_change as well as for Y_change are not statistically significant. Hence, we can conclude that the distribution of these two transformed variables is normal.

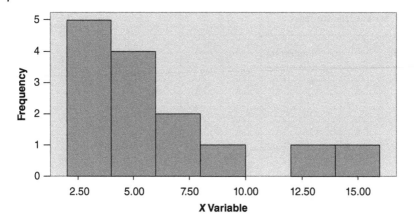

Figure 4.5 Histogram showing distribution of *X* (Maths scores).

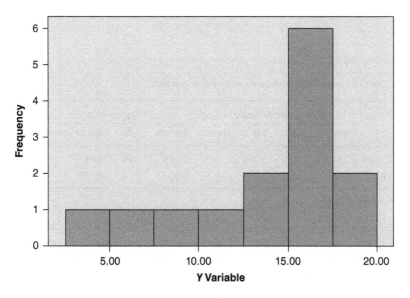

Figure 4.6 Histogram showing distribution of *X* (Science scores).

4.2.2 Randomness

Testing for randomness is a necessary assumption for the statistical analysis. The randomness is mostly related to the assumption that the data has been obtained from a random sample. Randomness of observations can be tested using Runs test. The null hypothesis for testing the randomness can be stated as "the data follows a random sequence."

Table 4.5 Transformed data.

Maths (X)	English (Y)	X_change log(X)	Y_change Y²
2	6	0.30	36
3	8	0.48	64
3	12	0.48	144
4	18	0.60	324
2	14	0.30	196
3	15	0.48	225
5	16	0.70	256
4	13	0.60	169
6	18	0.78	324
7	16	0.85	256
4	17	0.60	289
8	15	0.90	225
12	4	1.08	16
14	17	1.15	289

Table 4.6 Tests of normality for the transformed data on students' marks.

	Kolmogorov–Smirnov			Shapiro–Wilk		
	Statistics	df	Sig. (p-value)	Statistic	df	Sig. (p-value)
X_change	0.164	14	0.200	0.947	14	0.516
Y_change	0.164	14	0.200	0.911	14	0.163

4.2.2.1 Runs Test for Testing Randomness

Illustration 4.2 Consider a sample of 23 students drawn from a college to test whether the average weight of the college students is more than 130 lbs. The weights of the students so recorded are as follows:

Weight (lbs) 125 135 130 123 137 143 144 141 121 111 129 126 142 156 110 170 160 127 135 132 160 111 120

We shall now test whether the obtained sample is random using the Runs test in SPSS. Here, we are interested in testing the null hypothesis that the drawn sample is random against the alternative hypothesis that it is not.

After preparing the data file and using the sequence of commands **Analyze → Non Parametric Test → Runs Test,** the screen shown in Figure 4.7

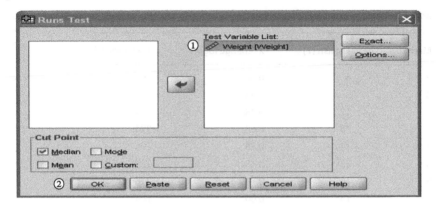

Figure 4.7 Screen showing options for Runs Test. Source: Reprint Courtesy of International Business Machines Corporation, © International Business Machines Corporation.

Table 4.7 Runs test for the data on weight.

	Weight
Test value[a]	132.00
Cases < test value	11
Cases ≥ test value	12
Total cases	23
Number of runs	11
z	−0.418
Asymptotic significance (two-tailed)	0.676

a) Median.

shall be obtained. Just by bringing the variable "Weight" from the left panel of the screen to the right panel and choosing the other options by default, the output as shown in Table 4.7 can be generated. The output shows that the absolute value of z statistic is not significant as its associated significance value is greater than 0.05, which means that the drawn sample is random.

4.2.3 Outliers

The first thing a researcher must do is to check the outlier in the data set. An outlier is an unusual data that exists in the data set. This kind of data may evolve due to wrong entry or due to extreme score existing in the data set. For instance, in entering data for height, most of the data may be within the range of 5.5–6.5 ft, but if by mistake 5.8 ft is entered as 58, then this becomes an

Table 4.8 Data showing the effect of outlier.

Data set A		Data set B	
X	Y	X	Y
1	5	1	5
2	4	2	4
3	3	3	3
4	1	4	1
5	2	5	2
		12	15
$r = -0.9$		$r = 0.81$	

unusual observation. Similarly, if any person in the sample has height as 7.2 ft, then it may be considered as an extreme observation. These outliers may distort the results completely. Hence, it is important that such data should be identified at the outset and appropriate action is taken before analyzing the data. Further, existence of the outliers in the data set may violate the assumption of normality.

As an example, let us see the effect of outlier in computing correlation. Consider the following two sets of data in Table 4.8. The correlation coefficient for the data set A is −0.9, but if one pair of outlier (12, 15) is introduced as shown in data set B, then the correlation becomes 0.81. Thus, we have seen that even one outlier may change the complexion of the correlation coefficient from −0.9 to 0.81.

4.2.3.1 Identifying Outliers with SPSS

We shall now show the procedure of identifying the outliers in SPSS. Generally, any observation outside mean $\pm$ 2SD or mean $\pm$ 3SD is considered as outliers. But the SPSS identifies the outliers using quartiles. In SPSS, an observation is identified as an outlier if it is below $Q_1 - (Q_3 - Q_1)/2$ or above $Q_3 + (Q_3 - Q_1)/2$. Its output provides the box plot and suggests the outliers by means of graphics.

Let us draw the box plot and identify outliers in the data set shown in Table 4.1. Using the same data file, click on the **Statistics** command in the screen shown in Figure 4.2. This will take you to the screen shown in Figure 4.8. Select the option "Outliers" and let other options be selected by default. Click on **Continue** and then **OK** to get the box plot in the output as shown in Figure 4.9. The box plot indicates that the first score in Maths and ninth score in the Science are outliers. The researchers may delete these outliers in the data for further analysis or correct them if it were due to typographical error.

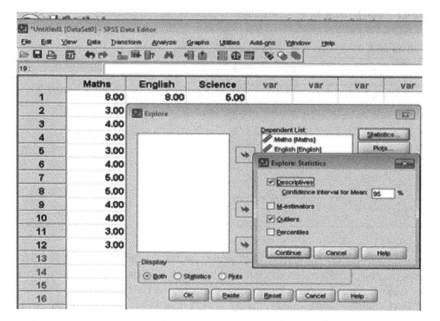

Figure 4.8 Screen showing option for selecting variables and identifying outliers. Source: Reprint Courtesy of International Business Machines Corporation, © International Business Machines Corporation.

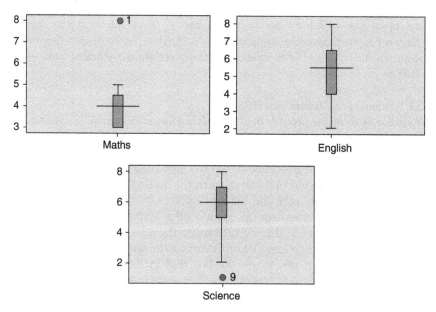

Figure 4.9 Box plot for all three groups of scores.

It is interesting to see that if you test the normality of data after removing the outliers, then it will become normal. Hence, by removing the outliers, the normality assumptions in using the parametric tests can also be fulfilled.

4.2.4   Homogeneity of Variances

The parametric test for comparing means of two groups assumes equal variances. The homogeneity of variances ensures that the samples are drawn from the populations having equal variance with respect to some criterion. The assumption for homogeneity is called as "Homoscedasticity," which is strongly influenced by non-normality. The departures from normality result in residuals, which account for the variances in sample data.

The normality assumption ensures that distribution of data is symmetric, while the equality of variances complements the same assuming homogeneous deviations from averages in subgroups: variance for sample 1 is assumed to be the same for sample 2. Homogeneity of variances can be tested using Box's M test, Levene's test, Bartlett's test, and graphical methods. The null hypothesis for equality of variances is stated as "equal variances are assumed/the data is homoscedastic."

4.2.4.1   Testing Homogeneity with Levene's Test

Levene's test is generally used for comparing the variability of two groups or more. Here, we test the null hypothesis of equality of variances against the alternative hypothesis that the variability differs. For null hypothesis to be retained, Levene's test should not be significant. In SPSS, this test is usually applied while comparing the means of the two groups. However, we shall show it independently.

Illustration 4.3   Let us consider the data of Illustration 4.1. We shall now test whether the variability of the two groups are same or not. We shall use the Levene's test in SPSS to test the following null hypothesis H_0: the variances of the two groups X and Y are same against the alternative hypothesis H_1: the variances are NOT equal (heterogeneity).

In order to apply the Levene's test in SPSS, the data file need to be prepared differently as shown in Table 4.9.

Using the sequence of commands **Analyze → Compare Means → Independent Samples *t* test** as shown in Figure 4.10, we shall get the screen as shown in Figure 4.11.

The steps involved in getting the output of Levene's test are shown in a sequential manner in Figure 4.11. By bringing the variables "Marks" and

Table 4.9 Data file for Levene's test.

Marks	Group	
2.0	1.0	Data of Group 1 (X)
3.0	1.0	
3.0	1.0	
4.0	1.0	
2.0	1.0	
3.0	1.0	
5.0	1.0	
4.0	1.0	
6.0	1.0	
7.0	1.0	
4.0	1.0	
8.0	1.0	
12.0	1.0	
14.0	1.0	
6.0	2.0	Data of Group 2 (Y)
8.0	2.0	
12.0	2.0	
18.0	2.0	
14.0	2.0	
15.0	2.0	
16.0	2.0	
13.0	2.0	
18.0	2.0	
16.0	2.0	
17.0	2.0	
15.0	2.0	
4.0	2.0	
17.0	2.0	

"Group" from the left panel to the respective places in the right panel and defining the group values as 1 and 2, the outputs of Levene's test can be generated, which is shown in Table 4.10. Since the significance value (p-value) of F is more than 0.05, the F-statistic is not significant. Thus, it may be concluded that the null hypothesis is retained and there is no evidence that the variability of the two groups differs.

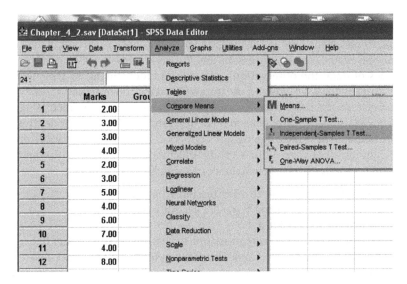

Figure 4.10 Screen showing commands for Levene's test in SPSS. Source: Reprint Courtesy of International Business Machines Corporation, © International Business Machines Corporation.

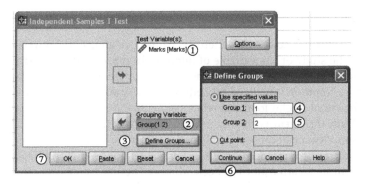

Figure 4.11 Screen showing commands for inputs in Levene's test. Source: Reprint Courtesy of International Business Machines Corporation, © International Business Machines Corporation.

Table 4.10 Results of Levene's test for equality of variances.

Variable	Levene's test	
	F	Sig. (p-value)
Marks	0.600	0.446

4.2.5   Independence of Observations

Various parametric tests require the independence assumption, including multiple linear regression, logistic regression, and Poisson regression. However, for certain types of data, the observations may not be independent; in repeated measures experimental designs and the time series data, the observations may be autocorrelated. The independence assumption can be tested by calculating intraclass correlations, Durbin-Watson (DW) test, and graphical methods. The DW test is available in "Linear Regression" option in SPSS, where the value of statistic is ignorable between -2 to $+2$ and perfect independence occurs when DW statistic is exactly "2." This could be done using SPSS as follows:

a) Choose from the menu Analyze > Regression > Linear Regression.
b) In the Linear Regression dialog box, click on Statistics.
c) Check the box for the DW test.
d) If the $-2 \leq$ DW test ≤ 2, then we can assume that there is no first-order linear autocorrelation in the observations. In another words, the observations are linearly independent.

We shall discuss this SPSS procedure while discussing linear regression later in this chapter.

4.2.6   Linearity

The linearity assumption states that there is a linear relationship between the variables of the study or the variable of interest. To preserve the linearity assumption, the first step is to choose variables that are theoretically linearly related. The most common visualized procedure to test linearity is the scatter plot, which is the visual display for linear relationship among variables. In linear regression analysis, the linearity in parameters is tested by the β coefficients. The statistical significance of these β coefficients depicts the linearity. In SPSS, choose the sequence of commands **Graphs → Legacy Dialogs → Scatter/Dot**. Look for the resulting scatter plot to check whether the linearity exists. The data points are presented in the shape of oval display linearity. If you see any other pattern for the data, it is most likely that the population from which your data is selected is not linear in terms of the variables you are studying. Thus, if you do not observe the oval shape indicative of linearity, your data violates the linearity assumption.

4.3   Assumptions in Hypothesis Testing Experiments

Several parametric and alternate nonparametric tests exist for hypotheses testing experiments. One of the common assumptions in using all parametric tests is that the data has been obtained from the population having normal distribution. Besides this, each parametric test has its own specific assumptions as well. If any of these assumptions is severely violated, the corresponding nonparametric test should be used. In this section, we shall discuss the assumptions in several commonly used parametric tests.

4.3.1 Comparing Means with *t*-Test

The simplest "*t*-test" is used for the comparison of location parameter (population mean) with a specified mean value. "One-sample *t*-test" is used for a small sample size ($n < 30$) when the population variance is unknown. In addition, the *t*-test statistic assumes an approximate normal distribution under the null hypothesis. The same method calculates the *t*-statistic as for the *z*-statistic (for large samples), with the only difference that the standard deviation here is replaced by the standard error of the estimate (the estimated standard deviation for $\overline{X}$). Also, the *t*-statistic retains a specific critical value for each level of significance and degrees of freedom (df) in published *t*-distribution table. The df, in general, is the sample size minus the number of the estimated parameters. So, the df for one population mean is $n-1$. In the following, the readers will learn about the three types of *t*-tests:

1) To compute the single-sample *t*-test and interpret the output when we have only one group and want to test the population's parameter against a hypothetical mean's value.
2) To compute the dependent (paired) *t*-test and interpret the output when we have two means for the same group. Both groups have the same people, or people are related/very similar, e.g. left hand–right hand, husband–wife, hospital patient and visitor.
3) To compute the independent two-sample *t*-test and interpret the output when we have two means and two groups, with no association between them.

4.3.2 One Sample *t*-Test

The one-sample *t*-test is typically used to compare a sample mean to a hypothetical population mean (given value μ_0). We use the following *t*-statistic for testing the hypothesis:

$$t = \frac{\overline{x} - \mu_0}{s/\sqrt{n}} \quad \text{with} \quad df = n - 1.$$

This test statistic follows *t*-distribution with $n-1$ degrees of freedom (df). The statistic *t* also follows normal distribution, but it is more flat at the top in comparison to the standard normal distribution. The flatness of the curve depends upon the sample size. We use *t*-statistic only when the sample is small

($n < 30$) and population variance is unknown. As per the central limit theorem, the distribution of t becomes normal if the sample is large ($n \geq 30$). Following are the assumptions in using the t-test:

1) Data should be free from outliers.
2) The data obtained on dependent variable must be metric (interval/ratio).
3) *Independence*: Each observation should be independent of each other.
4) *Randomness*: The data set should be randomly drawn.
5) *Normality*: The data obtained on dependent variable should follow a normal distribution or be approximately normal.

The outliers can be detected by box plot using the procedure shown in Section 4.2.3. Second assumption says that the t-test can be applied only for metric data (interval/ratio). But if the data is nonmetric (nominal/ordinal), one should use alternate nonparametric test. Assumption of independence of data can be tested by the procedure shown in Section 4.2.5. This will be deliberated by means of an example while discussing the assumptions in linear regression. In the following sections, we shall show the procedure of testing randomness and normality in one-sample t-test and the repercussion thereof if the assumption breaks down.

4.3.2.1 Testing Assumption of Randomness

Randomness of data can be tested by Runs test. We shall show the procedure of testing randomness by means of Illustration 4.4.

Illustration 4.4 A sample of 16 subjects was selected for the study, whose anxiety scores are shown in Table 4.11. Let us see whether this sample can be considered to be randomly drawn at 5% level.

In this illustration, we need to test the following hypotheses at a significance level of 0.05.

H_0: Sample is random.
H_1: Sample is biased.

We shall use the Runs test to test the null hypothesis. The test statistic shall be the number of Runs (r_c). Let us see how we compute the runs. A score larger

Table 4.11 Anxiety scores.

50	54	32	45	48	87	80	86	85	92	94	25	60	65	50	42

than the previous is a given "+" sign and smaller than the previous is a given "–" sign. The first score has been given a "–" sign.

50	54	32	45	48	87	80	86	85	92	94	25	60	65	50	42
–	+	–	+	+	+	–	+	–	+	+	–	+	+	–	–

The runs have been made by keeping the consecutive signs "+" and "–" together. Thus, the total number of runs are as follows:

–	+	–	+++	–	+	–	++	–	++	––
1	2	3	4	5	6	7	8	9	10	11

Here,

n_1 = number of "–" sign = 7,
n_2 = number of "+" sign = 9,
r = total number of runs = 11.

From Table A.3, the lowest and highest critical values of the runs (r_c) for $n_1 = 7$ and $n_2 = 9$ at 5% level are 4 and 14.

Since observed number of runs r (=11) lies between 4 and 14, the null hypothesis H_0 is not rejected at 5% level. Since the null hypothesis has not been rejected, it may be concluded that the sample drawn was random.

4.3.2.2 Testing Normality Assumption in t-Test

We shall show the procedure of testing normality in one-sample t-test using SPSS with the help of Illustration 4.5.

Illustration 4.5 Consider the marks obtained by the students in mathematics as shown in Table 4.12. We want to test the hypothesis that the sample has been selected from a population having a mean of 8. In other words, we want to test the null hypothesis H_0: $\mu = 8$ against the alternative hypothesis H_1: $\mu \neq 8$.

Since population standard deviation is unknown, t-test can be used to test the null hypothesis. But one of the assumptions in using the t-test is that the sample has been drawn from a normal population.

Let us first check the distribution of the sample with Shapiro–Wilk test using SPSS as discussed in Section 4.2.1. Table 4.13 shows that the Shapiro test for the data on mathematics is significant as the significance value associated with this test is 0.002, which is less than 0.01. This indicates that the normality assumption is violated. Let us not bother about violation of this assumption and apply t-test using SPSS first.

Table 4.12 Students marks in mathematics.

Marks (X)
3.0
4.0
4.0
5.0
6.0
5.0
6.0
4.0
5.0
7.0
6.0
7.0
4.0
5.0
7.0
12.0
6.0
12.0
15.0
14.0

Table 4.13 Tests of normality for the data on students' marks.

	Kolmogorov–Smirnov			Shapiro–Wilk		
	Statistics	df	Sig. (*p*-value)	Statistic	df	Sig. (*p*-value)
Mathematics (X)	0.264	20	0.001	0.823	20	0.002[a]

a) Significant at 5% level.

t-test After making the data file in SPSS and using the sequence of commands **Analyze → Compare Means → One Sample *t* test,** we shall get the screen as shown in Figure 4.12.

By transferring the variable "Marks" from the left panel to the "Test Variable(s)" section and entering population mean, 8, in the "Test Value"

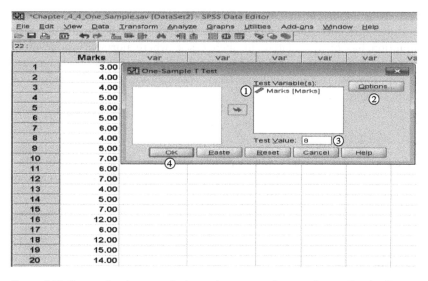

Figure 4.12 Screen showing steps for inputs in one-sample *t* test. Source: Reprint Courtesy of International Business Machines Corporation, © International Business Machines Corporation.

Table 4.14 Descriptive statistics.

	N	Mean	Standard deviation	Standard error mean
Marks	20	6.8500	3.51351	0.78564

Table 4.15 T-table for the data on marks in mathematics.

t	df	Sig. (p-value) (two-tailed)	Mean difference
−1.464	19	0.160	−1.15000

Test value = 8

section, the outputs of the *t* test can be generated. These outputs are shown in Tables 4.14 and 4.15.

It can be seen from Table 4.15 that the absolute value of *t*-statistic is not significant as its significance value is 0.160, which is more than 0.05; hence, the null hypothesis is retained. In other words, there is no evidence that the sample has been drawn from a population whose mean is not 8.

4.3.2.3 What if the Normality Assumption Is Violated?

The above inference about null hypothesis is doubtful because normality assumption has been violated. In such cases, where normality is severely violated, nonparametric test should be applied. For one-sample test, sign test is the appropriate test in nonparametric. We shall now check the results of hypothesis testing with this test.

4.3.3 Sign Test

We shall now check the null hypothesis H_0: $\mu = 8$ against the alternative hypothesis H_1: $\mu \neq 8$ using the sign test. Let us replace each score greater than 8 with a plus sign and less than 8 with a minus sign:

Marks	3	4	4	5	6	5	6	4	5	7	6	7	4	5	7	12	6	12	15	14
Sign	−	−	−	−	−	−	−	−	−	−	−	−	−	−	−	+	−	+	+	+

Number of "+" sign = 4
Number of "−" sign = 16
N = number of scores = 20

here, the statistic, X = number of fewer sign = 4. From Table A.4, the critical value for the statistics X (fewer sign) for $n = 20$ at 5% level is equal to 5. Since the value of X (=4) is less than the critical value 5, the null hypothesis is rejected, and therefore, it may be concluded that the mean of the distribution is not 8.

We have seen that t-test does not reject the null hypothesis due to violation of normality assumption, but the sign test rejects the null hypothesis. Thus, in such situations where normality assumption is severely violated, t-test should not be used, instead alternate nonparametric test should be preferred to detect the effect.

4.3.4 Paired t-Test

Paired t-test is used to compare two related means (μ_1 and μ_2), which mostly comes from a repeated measures design. In other words, data is collected by two measures from each observation, e.g. before and after. Where H_0 is the tested null hypothesis for testing if the difference between two dependent means equals to a specific value (zero could be one of the possible values)? For example, a researcher wants to test if the changes in the weight before and after a family therapy intervention are significantly different from zero. In paired t-test, all assumptions are applicable on the difference of pre and post data. Following are the assumptions in paired t-test:

1) The scores obtained on dependent variable (pre–post) should be free from outliers.

2) The data obtained on dependent variable must be continuous (interval/ratio).
3) *Independence*: Each pair of observations on dependent variable (pre–post) should be independent of each other.
4) *Randomness*: The data set should be randomly drawn.
5) *Normality*: The data obtained on dependent variable (pre–post) should follow a normal or approximately normal distribution.

We shall show how the assumption of normality is tested and the repercussion thereof if it is violated by means of the following illustration. The readers should note that the t-test is robust to any violations of the assumptions of normality, particularly for moderate ($n > 30$) and larger sample sizes.

Illustration 4.6   Consider an experiment in which an exercise programme was launched to know whether it increases the hemoglobin contents of the participants. Fourteen participants took part in the three-week exercise programme. Their hemoglobin content was measured before and after the programme, which are shown in Table 4.16.

Here, we need to test the null hypothesis that there is no difference in the mean hemoglobin scores after and before the exercise programme against the alternative hypothesis that the program is effective. Since our data is metric, a paired t-test would be appropriate to test the hypothesis. But one of the assumptions for this test is that the difference of post and pre data must be normally distributed. Hence, let us first test the normality of the difference of scores using the procedure discussed in Section 4.2.1.

After applying the Shapiro test in SPSS, the output so generated is shown in Table 4.17. This table clearly indicates that the Shapiro test for the data on

Table 4.16 Hemoglobin contents of the participants before and after the exercise programme.

Before (X)	6	7.2	6	8.2	9.5	10.5	8.5	7.3	8.6	12	10.8	7.8	8.2	9.1
After (Y)	6.5	7.1	6.2	8.3	12.4	10.6	8.7	7.5	11.2	11	10.9	7.9	8.5	9.5
Difference ($Y - X$)	0.5	−0.1	0.2	0.1	2.9	0.1	0.2	0.2	2.6	−1	0.1	0.1	0.3	0.4

Table 4.17 Tests of normality for the data on hemoglobin.

	Kolmogorov–Smirnov			Shapiro–Wilk		
	Statistics	df	Sig. (p-value)	Statistic	df	Sig. (p-value)
Difference (D)	0.346	14	0.000[a]	0.713	14	0.001[a]

a) Significant at 5% level.

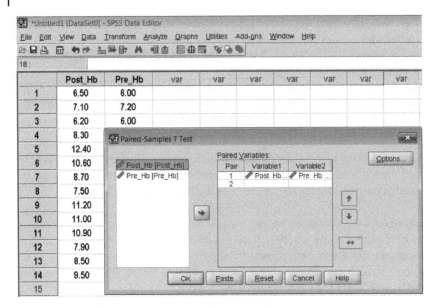

Figure 4.13 Screen showing option for paired t test. Source: Reprint Courtesy of International Business Machines Corporation, © International Business Machines Corporation.

Table 4.18 Paired t-test for the data on hemoglobin.

Paired differences			95% CI of the difference					
	Mean	SD	SE of mean	Lower	Upper	t	df	Sig. (p-value) (two-tailed)
Post_Hb–Pre_Hb	0.47	1.03	0.27	−0.12	1.07	1.72	13	0.110

difference is significant ($p < 0.01$); hence, it may be concluded that the normality assumption is violated in this case. Let us see what happens if we don't bother about normality and apply the t-test using the SPSS. After using the sequence of commands **Analyze → Compare Means → Paired Samples t test**, we get the screen as shown in Figure 4.13. On this screen, transferring the Post_Hb and Pre_Hb variables from the left panel to the right panel and pressing on **OK** will generate the output of paired t-test as shown in Table 4.18. The output reveals that the t-test is not significant; hence, the null hypothesis cannot be rejected and it may be concluded that the exercise programme is not effective in improving the hemoglobin.

4.3.4.1 Effect of Violating Normality Assumption in Paired *t*-Test

The findings in Illustration 4.5 have been obtained under the violation of normality assumption. Here, normality assumption is severely violated; hence, we shall try the nonparametric test for testing the effect. For paired samples, the appropriate nonparametric test is Rank test. Let us see what conclusion we get if we use this test.

4.3.5 Rank Test

In order to apply the rank test for testing the effectiveness of the programme in increasing the hemoglobin, we shall first compute the difference (D) between post and pre data for each pair. If the difference is positive, a plus (+) sign is marked for that pair, and if the difference is negative, a negative (−) sign is marked. If the difference is zero, that pair of scores is deleted from the study. These computations have been shown in Table 4.19.

One can see that number of plus (+) sign = 12, number of minus (−) sign = 2, number of paired scores $n = 14$, and the test statistics $X =$ number of fewer sign = 2.

From Table A.4, for $n = 14$ and at 5% level, the critical value of the fewer sign is 3. Since the test statistic X (=2) is less than the critical value 3, the null hypothesis is rejected. Thus, it may be concluded that the exercise program

Table 4.19 Data on hemoglobin and computation in sign test.

S. No.	Pre_Hb	Post_Hb	Difference (*D*)	Sign
1	6	6.5	0.5	+
2	7.2	7.1	−0.1	−
3	6	6.2	0.2	+
4	8.2	8.3	0.1	+
5	9.5	12.4	2.9	+
6	10.5	10.6	0.1	+
7	8.5	8.7	0.2	+
8	7.3	7.5	0.2	+
9	8.6	11.2	2.6	+
10	12	11	−1	−
11	10.8	10.9	0.1	+
12	7.8	7.9	0.1	+
13	8.2	8.5	0.3	+
14	9.1	9.5	0.4	+

is effective in enhancing the hemoglobin. We have seen that if the normality assumption is severely violated, the result of the study is completely reversed. In other words, effect could be seen using the nonparametric test, whereas due to violation of normality assumption, paired *t*-test was unable to detect the effect though it existed.

4.3.6 Independent Two-Sample *t*-Test

One of the most widely and commonly used parametric test is the **independent two-sample *t*-test**. It is utilized for comparing differences between two separate groups/populations. In psychology, these groups are often collected by randomly assigning research participants to conditions. However, this test may also be performed to explore differences in naturally occurring groups. For example, comparing between male and female on emotional intelligence, the independent two-sample *t*-test can be used to test the significance of difference. Following are the assumptions of a two-sample *t*-test:

1) Both the data set should be free from outliers.
2) The data obtained on dependent variable in each group must be continuous (interval/ratio).
3) *Independence*: Each subject should get only one treatment. In other words, the subjects in both the groups should be different.
4) *Randomness*: The data in each group should be obtained randomly.
5) *Normality*: The data obtained on dependent variable in each group should follow a normal distribution or be approximately normal.
6) Variability of the data in both the groups should be same.

4.3.6.1 Two-Sample *t*-Test with SPSS and Testing Assumptions
In this section, readers will know how to check the assumptions of normality and equal variance and to apply *t*-test using the SPSS by illustration.

Illustration 4.7 The memory retention scores of male and female are shown in Table 4.20. It is desired to test whether memory of a person is gender specific. To test the null hypothesis H_0: $\mu_{male} = \mu_{female}$ against the alternative hypothesis H_0: $\mu_{male} \neq \mu_{female}$, we shall explore an appropriate test.

Since the data is metric (quantitative), for testing the null hypothesis, independent-sample *t*-test shall be used. For this test, we require the assumption of normality and homogeneity. Let us check these assumptions first. Using the procedure as discussed in Section 4.2.1, we can obtain the results for normality as shown in Table 4.21.

It can be seen that the Shapiro test for the memory data of male is significant; hence, it may be concluded that the normality assumption is violated. Let us not bother about this assumption and proceed further in applying the independent-samples *t*-test for testing our hypothesis.

Table 4.20 Memory scores.

Male	Female
3	7
4	8
4	6
5	5
6	8
5	9
6	9
4	8
5	8
7	7
6	6
7	11
4	12
5	9
11	9
12	12
6	10
12	6
15	7
14	5

Table 4.21 Tests of normality for the data on memory.

	Kolmogorov–Smirnov			Shapiro–Wilk		
	Statistics	df	Sig. (p-value)	Statistic	df	Sig. (p-value)
Male	0.264	20	0.001[a]	0.823	20	0.002[a]
Female	0.132	20	0.200	0.946	20	0.308

a) Significant at 5% level.

In order to test the homogeneity assumption, we shall apply Levene's test. The result of this test is simultaneously obtained while applying independent-samples t-test in SPSS. The procedure of preparing the data file is slightly different in this case, the format of which is shown in Table 4.22. Here, "Group" and "Memory Score" have been defined as nominal and scale variables, respectively. Group has been coded as 1 for Male and 2 for Female.

Table 4.22 Data file in SPSS for independent-sample *t*-test.

	Group	Memory score
Male	1.0	3.0
	1.0	4.0
	1.0	4.0
	1.0	5.0
	1.0	6.0
	1.0	5.0
	1.0	6.0
	1.0	4.0
	1.0	5.0
	1.0	7.0
	1.0	6.0
	1.0	7.0
	1.0	4.0
	1.0	5.0
	1.0	11.0
	1.0	12.0
	1.0	6.0
	1.0	12.0
	1.0	15.0
	1.0	14.0
Female	2.0	7.0
	2.0	8.0
	2.0	6.0
	2.0	5.0
	2.0	8.0
	2.0	9.0
	2.0	9.0
	2.0	8.0
	2.0	8.0
	2.0	7.0
	2.0	6.0
	2.0	11.0
	2.0	12.0
	2.0	9.0
	2.0	9.0
	2.0	12.0
	2.0	10.0
	2.0	6.0
	2.0	7.0
	2.0	5.0

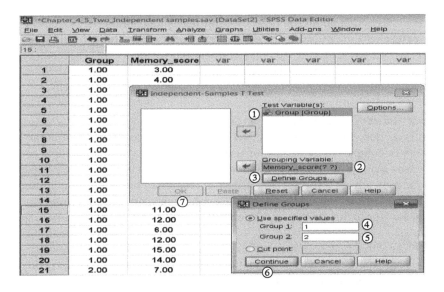

Figure 4.14 Screen showing options for independent-samples t-test. Source: Reprint Courtesy of International Business Machines Corporation, © International Business Machines Corporation.

Table 4.23 Independent-samples t-test for the data on memory.

	Levene's test for equality of variances		t-Test for equality of means				
	F	Sig. (p-value)	t	df	Sig. (p-value)[a]	Mean difference	SE of difference
Equal variances assumed	5.207	0.028	−1.122	38	0.269	−1.05	0.936
Equal variances not assumed			−1.122	30.197	0.271	−1.05	0.936

a) Two-tailed test.

After preparing the data file and using the sequence of commands **Analyze → Compare Means → Independent-Samples t test** in SPSS, the screen shown in Figure 4.14 can be obtained.

By shifting "Group" and "Memory_score" variables from the left panel to their respective places in the right panel as shown in the screen and by specifying the values of Group 1 as 1 and Group 2 as 2, the outputs of independent-samples t test can be obtained, which is shown in Table 4.23. The output also includes the results of Levene's test. Since F-value ($p < 0.05$) is significant, it may be concluded that the variances of both the groups are not same. In other words, homogeneity assumption is violated.

Let us not bother about the violation of normality and homogeneity assumptions and check the results of the t-test. It can be seen from Table 4.23 that the t-test is not significant ($p > 0.05$); hence, there is no evidence that the memory of male and female differs. The readers must note that this finding has been obtained under the violation of normality and homogeneity assumptions.

4.3.6.2 Effect of Violating Assumption of Homogeneity

In this section, we shall see the effect of violating the assumption of homogeneity in comparing the means of two groups. In the above illustration, we have seen that the homogeneity and normality assumptions have been violated. We have already seen the effect of extreme violation of normality assumption on findings. Now we shall show that if the homogeneity assumption is violated, how the parametric test (independent-sample t-test) may give the wrong findings.

We know that in case of metric data, the appropriate test for comparing the means of two groups under the situations when the assumptions of normality and homogeneity are violated is Mann–Whitney test. Let us see what happens when we use this nonparametric test to test our desired hypothesis. This test can be applied in SPSS using the same data file we have prepared for the t-test. Using the sequence of commands **Analyze → Nonparametric Tests → 2 Independent Samples**, the screen shown in Figure 4.15 can be obtained.

By shifting "Group" and "Memory_score" variables from the left panel to their respective places in the right panel as shown in the screen, by specifying the values of Group 1 as 1 and Group 2 as 2, and by ensuring the selection of option for Mann–Whitney test (already selected by default), the outputs of the

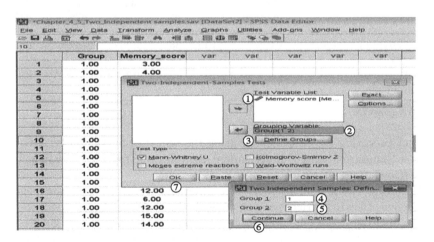

Figure 4.15 Screen showing options for Mann–Whitney test. Source: Reprint Courtesy of International Business Machines Corporation, © International Business Machines Corporation.

Table 4.24 Test statistics for Mann–Whitney test.

	Memory score
Mann–Whitney U	126.500
Wilcoxon W	336.500
Z	−2.003
Asymptotic significance (two-tailed)	0.045

test can be obtained, which is shown in Table 4.24. Since z-value is significant ($p < 0.05$), it is concluded that the mean memory score of male and female differs significantly.

We have seen that without bothering the assumptions of normality and homogeneity, independent-samples t-test suggests that there is no difference between the mean memory scores of male and female, but once we applied the nonparametric Mann–Whitney test, the difference becomes significant. Of course, in this illustration, the assumptions are severely violated due to which there is a discrepancy in the findings. In experimental studies, the effect may exist but may not be detected if parametric test is used without bothering to check the assumptions.

Remark Homogeneity assumptions can also be tested by the thumb rule. If ratio of the two standard deviations is less than 2, we can assume that the homogeneity assumption is not violated. In other words, SD_{max}/SD_{min} should always be less than 2.

4.4 *F*-test For Comparing Variability

In the previous sections, we covered all the three t-tests for estimating and testing the population's mean. The F-test is different from t-tests, where researchers/practitioners utilize it for testing whether there are any differences in the variances within the samples. F-test is used for testing equality of variances before using the t-test. In cases where there are only two means to compare, the t-test and the F-test are equivalent and generate the same results. Consequently, this is why F-test is considered an extension of the independent two-sample t-test.

This analysis can only be performed on numerical data (data with quantitative value). For a continuous variable, the F-test could be used in analyzing variance to see if three or more samples come from populations with equal means. Analysis of variance (ANOVA) is the appropriate statistical technique for testing differences among three or more means (treatments or groups). However, it seems strange that the technique is called ANOVA instead of analysis of means. The readers will see that the name ANOVA is suitable because hypothesis testing about means are made by analyzing variance. Hence, ANOVA is utilized for testing general instead of specific differences among means based on the F-test.

4.4.1 Analysis of Variance (ANOVA)

In the ANOVA, the response variable is the variable we are comparing. The factor variable is the categorical variable being used to define the groups. We will assume k samples (groups or treatments). We utilize the ANOVA hypothesis test to detect significant differences in population's means for more than two populations. There are several types of ANOVA, depending on the number of the categorical variables. In one-way ANOVA (where there is only one categorical variable), each value is classified in exactly one way. If we have two categorical variables, then it is named as two-way ANOVA, and so on. Example for ANOVA includes comparisons by sex, color, race, political party, and so on. Suppose we want to test the differences in means among three groups as in Figure 4.16.

If we use the t-test separately between two means, we will increase Type I error (α), where $\alpha = 0.15$ ($0.05 \times 3 = 0.15$) approximately and confidence level $(1-\alpha)^3 = (1-0.05)^3 = (0.95)^3 \cong 0.85$. Therefore, we use a technique called ANOVA.

The basic idea behind the ANOVA is partitioning the total variation of the data into two sources:

1) The effect of the factor/treatment/group, which is the variability between the groups, SS(B).
2) The effect of the random errors or effect of the other variables, which are not in the study. It is called the variation within the groups, SS(W).

Let's recall that the

- Variation is known as the sum of the squares of the deviations between an observation and the mean of all the observations in the study.
- SS refers to the sum of squares and often is followed by a variable in parentheses such as SS(B) or SS(W) to clarify which sum of squares we're talking about.

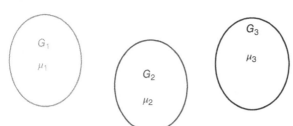

Figure 4.16 Three different groups/treatments with their means.

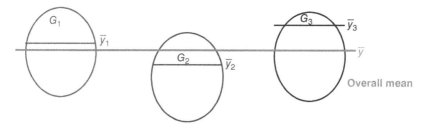

Figure 4.17 Three different groups/treatments with their means and overall mean.

In ANOVA, we study the differences in means among the groups and overall mean. In addition, we study the differences in means within the groups as in Figure 4.17.

4.4.2 ANOVA Assumptions

There are several assumptions in ANOVA that should be investigated before proceeding to the inferential analysis. The main assumptions are

- Independence
- Normality
- Homogeneity of variances (e.g. homoscedasticity)

4.4.2.1 Checking Assumptions Using SPSS

The Independence, Normality, and Homogeneity assumptions can be tested in SPSS using some in-built tests as discussed before. The readers should be aware that the Independence assumption for ANOVA is more theoretical, assuming the three groups of respondents to be independent of each other (referring to the random selection of respondents in each group from the population, i.e. a random sample). In other words, each subject should get one and only one treatment.

Illustration 4.8   A study was conducted for the IQ scores of students (out of 60) from three groups of undergraduate students (15 in each group) of different disciplines: Physics, Maths, and Chemistry. The raw data is shown in Table 4.25.

Let us make the data file in SPSS by defining three variables: Physics, Maths, and Science as Scale measure. The data file will look like Figure 4.18. The readers may refer the procedure of making data file in SPSS as discussed in Chapter 2.

After making the data file for testing the normality using measures of skewness and kurtosis, follow the given sequence from SPSS menu: **Analyze → Descriptive Statistics → Descriptives** (Figure 4.19).

After clicking on the **Descriptives,** choose the variables in the **Variable(s)** to be tested for skewness and kurtosis. By clicking on **Options** tab, a window will

Table 4.25 IQ scores of students.

Physics	Maths	Chemistry
44	36	52
40	40	50
44	37	51
39	35	52
25	39	45
37	40	49
31	36	47
40	38	46
22	24	47
34	27	47
39	29	46
20	24	45
39	45	50
42	44	47
41	44	49

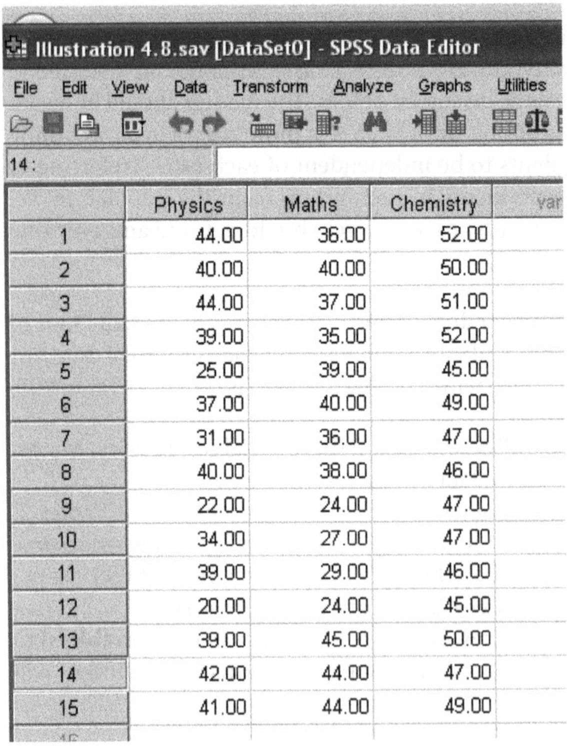

	Physics	Maths	Chemistry	var
1	44.00	36.00	52.00	
2	40.00	40.00	50.00	
3	44.00	37.00	51.00	
4	39.00	35.00	52.00	
5	25.00	39.00	45.00	
6	37.00	40.00	49.00	
7	31.00	36.00	47.00	
8	40.00	38.00	46.00	
9	22.00	24.00	47.00	
10	34.00	27.00	47.00	
11	39.00	29.00	46.00	
12	20.00	24.00	45.00	
13	39.00	45.00	50.00	
14	42.00	44.00	47.00	
15	41.00	44.00	49.00	

Figure 4.18 Data file in SPSS for testing ANOVA assumptions. Source: Reprint Courtesy of International Business Machines Corporation, © International Business Machines Corporation.

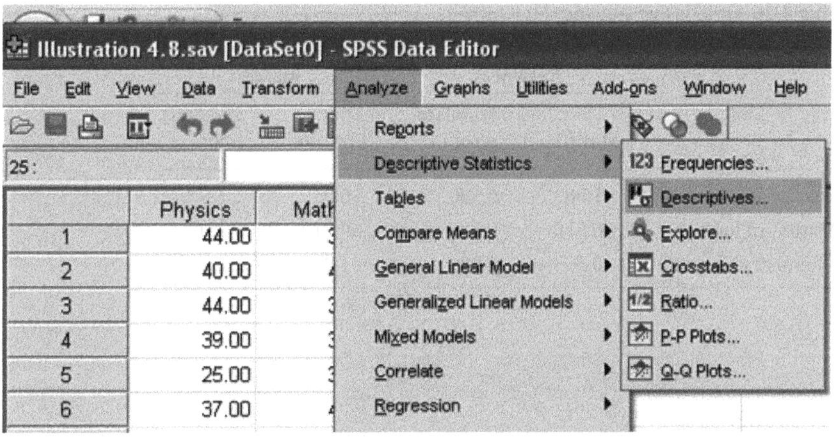

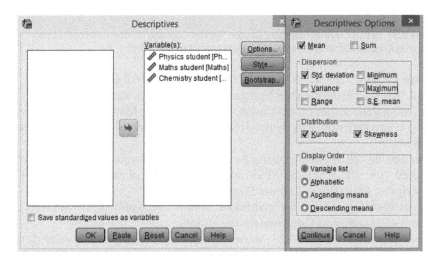

Figure 4.19 Path for testing Skewness and Kurtosis. Source: Reprint Courtesy of International Business Machines Corporation, © International Business Machines Corporation.

Figure 4.20 Procedure for testing Skewness and Kurtosis. Source: Reprint Courtesy of International Business Machines Corporation, © International Business Machines Corporation.

appear providing the options for various descriptive measures for the data. Check on for "Skewness" and "Kurtosis" as shown in Figure 4.20. Click on **Continue** and **OK** on the main screen.

Following the rule of thumbs, the skewness can be interpreted for testing the normality assumption. Since the absolute value of skewness for all three

Table 4.26 Descriptive statistics for IQ scores.

| | Skewness | | Kurtosis | |
	Statistic	Standard error	Statistic	Standard error
Physics student	−1.083	0.580	−0.049	1.121
Maths student	−0.578	0.580	−0.710	1.121
Chemistry student	0.303	0.580	−1.220	1.121

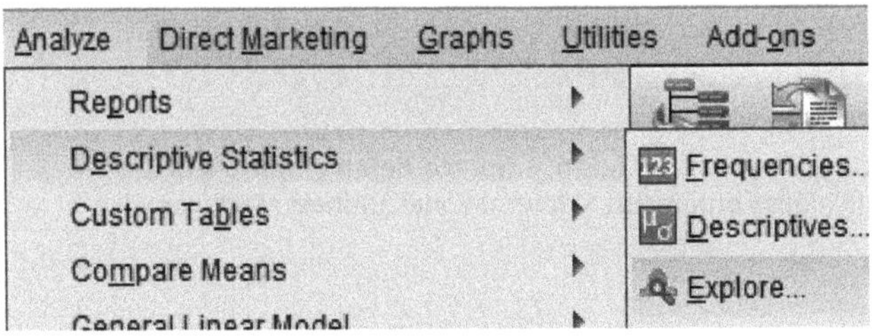

Figure 4.21 Path for testing normality. Source: Reprint Courtesy of International Business Machines Corporation, © International Business Machines Corporation.

groups in Table 4.26 is less than two times of its standard error, i.e. less than $(2 \times 0.580) = 1.16$, the data can be regarded as free from a significant skewness. According to the other rule of thumb, the IQ scores for Maths and Chemistry students are approximately symmetric, while moderately skewed for Physics students.

For testing the normality using formal tests, the Shapiro–Wilk test is to be used ($n < 50$). From the SPSS menu, follow the sequence: **Analyze →Descriptive Statistics → Explore** (Figure 4.21).

From **Plots** tab, choose the mentioned option "Normality plots with tests" (Figure 4.22) and press **Continue** for the output.

The results in Table 4.27 show the test statistics and p-value for Kolmogorov–Smirnov and Shapiro–Wilk tests. Due to a small sample size, we will be looking at Shapiro–Wilk test. A significant p-value ($p = 0.015 < 0.05$) depicts that the IQ scores for Physics students are not normally distributed. However, the scores for Maths and Chemistry students conform to the normal distribution (p-value $= > 0.05$).

The results from **Explore** option also construct Q–Q plot for each selected variable by default. Q–Q plot is a graphical way of showing the normality of the data. It compares two probability distributions by plotting their quintiles against each other. If the distribution of the observed sample is similar to that of the standard normal distribution, the points in the Q–Q plot will lie on the

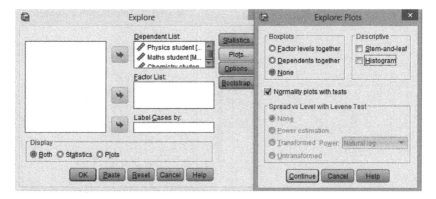

Figure 4.22 Procedure for testing normality. Source: Reprint Courtesy of International Business Machines Corporation, © International Business Machines Corporation.

Table 4.27 Tests for normality.

	Kolmogorov–Smirnov			Shapiro–Wilk		
	Statistic	df	Sig. (p-value)	Statistic	df	Sig. (p-value)
Physics_IQ	0.259	15	0.008	0.846	15	0.015
Maths_IQ	0.184	15	0.185	0.910	15	0.135
Chemistry_IQ	0.225	15	0.040	0.917	15	0.174

line. The line indicates that the value of your sample should fall on it or close by if the data follows normal distribution. Deviation of these points from the line indicates non-normality. The normal $Q–Q$ plot (Figure 4.23) shows that the IQ scores for Maths and Chemistry are aligned with the normal quintiles line but the IQ scores for Physics students violates the normality assumption.

The homogeneity assumption can be checked using the Levene's test for equality of variances in SPSS ANOVA setup. To start with testing this assumption, use SPSS data file of Figure 4.18. To go into the SPSS ANOVA setup, follow the sequence **Analyze → General Linear Model → Univariate** to get the screen as shown in Figure 4.24. Choose "IQ score" in "Dependent Variable" and "group number" in the "Fixed Factor(s)" sections. To directly go for Levene's test in the setup, click on **Options** tab in the **Univariate** window. Choose "Homogeneity Tests" and click on **Continue** for the results.

Since the p-value ($p = 0.005$) for Levene's test of equality of error variances in Table 4.28 is less than the significance value of 0.05, the hypothesis of equal variances is rejected. This result indicates that the error variances for IQ score are not equal across the area of specialization (groups).

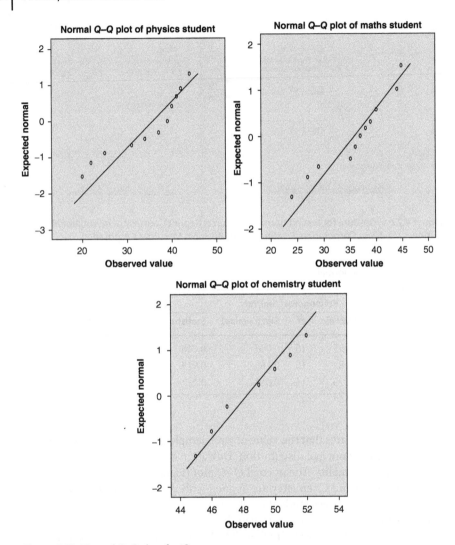

Figure 4.23 Normal Q–Q plots for IQ scores.

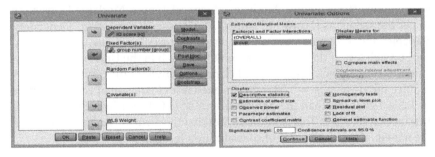

Figure 4.24 Procedure for Levene's test. Source: Reprint Courtesy of International Business Machines Corporation, © International Business Machines Corporation.

Table 4.28 Levene's test of equality of error variances.

Dependent variable: IQ score			
F	df1	df2	Sig. (p-value)
5.932	2	42	0.005

Tests the null hypothesis that the error variance of the dependent variable is equal across groups.

4.4.3 One-Way ANOVA Using SPSS

Assume that we want to compare k groups of drugs to recovery time. The null hypothesis is that all of the population means are equal (assuming that k drugs are the same), and the alternative is that not all of the means are equal. In other words, the purpose of the ANOVA table is to determine whether at least two of the populations have significantly different population mean values. Quite often, though the two hypotheses are given as

$$H_0 : \mu_1 = \mu_2 = \mu_3 = \cdots = \mu_k.$$
$$H_a : \text{At least one is different.}$$

Estimate the common value of the variance (s^2):

1) The effect of the factor/treatment/group, which is the variability between the groups, SS(B).
2) The effect of the random errors or effect of the other variables, which are not in the study. It is called the variation within the groups, SS(W).

Simply, the F-test equals the ratio of the two variance estimates (dividing the between group variance by the within group variance). Also, recall the mean square (MS), MS between group variation, and the MS within group variation.

$$F = \frac{\text{MSB}}{\text{MSW}}.$$

An excessively large F-test statistic is evidence against equal population means.

If the F-test result shows that null hypothesis cannot be rejected, then the research should stop and the conclusion is written that there is no statistically significant difference among the groups. However, if the results show that the null hypotheses should be rejected, then it indicates that the difference among the groups is statistically significant. Consequently, the researchers should follow-up the analysis by doing some follow-up tests. In other words, the post-hoc comparisons for ANOVA should be performed if and only if you reject H_0 in the F-test for ANOVA. The post-hoc comparison of each pair of groups is performed to identify which groups have different population means. Pairwise comparisons between group means are also referred to as contrasts.

Let us assume that there is a factor or treatment with three levels. Then the researcher has to perform three pairwise comparisons to compare among the groups' means (i.e. groups 1 and 2, groups 1 and 3, and groups 2 and 3).

Different types of post-hoc tests are used for doing the multiple comparisons and protecting Type I error (α). Some of them are discussed below.

- *Bonferroni test:* It uses t-tests to do pairwise comparisons between group means but controls the overall error rate (α) for each test to the experiment-wise error rate divided by the total number of tests. Hence, the observed significance level is adjusted for the fact that multiple comparisons are being made.
- *Tukey test:* To make all of the pairwise comparisons among groups, we utilize the studentized range statistic. It sets the error rate at the experiment-wise error rate for the collection of all the pairwise comparisons.
- *Scheffe test:* It implements simultaneous joint pairwise comparisons for all possible pairwise combinations. It uses the sampling distribution as F-distribution.
- *Dunnett's C:* It is a pairwise comparison test based on the studentized range. This test is appropriate when the variances are not equal.
- *Tamhane's T2:* It is used when the variances are not equal and is a conservative pairwise comparison test that relies on the t-test.
- *Games-Howell:* It is also used when the variances are not equal and is a liberal pairwise comparison test.

One-way ANOVA in SPSS can be illustrated using the SPSS data file in Figure 4.18. One-way ANOVA can be carried out in two ways in SPSS: by choosing either **General Linear Model (GLM)** or **Compare Means** (used in this book). The GLM however provides more options to customize the model to be tested under ANOVA. The sequence for both methods is given as follows:

Method 1: Analyze → Univariate → General Linear Model (GLM)
Method 2: Analyze → Compare Means → One-Way ANOVA

The analysis using Method 2 is simpler to use. From the SPSS menu, go from **Analyze** to **Compare Means** and choose the **One-Way ANOVA** as shown in Figure 4.25.

Choose the "IQ Score" (continuous variable) in "Dependent List" and the "group" (nominal variable) in the "Factor" (Figure 4.26). From the **Options** tab, click for "Means Plot," and click on **Continue** for the results.

The Output Interpretation The results from an ANOVA test are shown in Table 4.29. Since the p-value associated with F is significant (p-value $= 0.000 <$ 0.05), the null hypothesis of equality of means may be rejected. This means

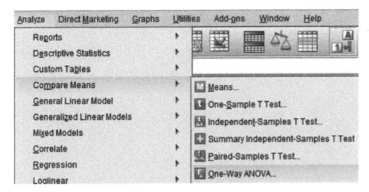

Figure 4.25 Path for One-Way ANOVA. Source: Reprint Courtesy of International Business Machines Corporation, © International Business Machines Corporation.

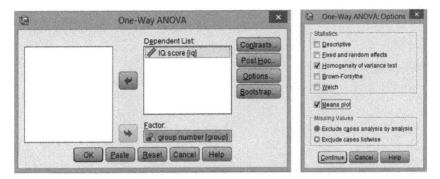

Figure 4.26 Procedure for One-Way ANOVA. Source: Reprint Courtesy of International Business Machines Corporation, © International Business Machines Corporation.

Table 4.29 The analysis of variance (ANOVA) table.

IQ score					
	Sum of squares	df	Mean square	F	Sig. (p-value)
Between groups	1529.378	2	764.689	20.016	0.000
Within groups	1604.533	42	38.203		
Total	3133.911	44			

that at least one of the groups means is different. The means plot in Figure 4.27 shows that the Chemistry students outperformed others in their IQ score. However, no formal results are provided regarding the group causing this difference between IQ scores. A post-hoc analysis is to be carried out to serve the purpose.

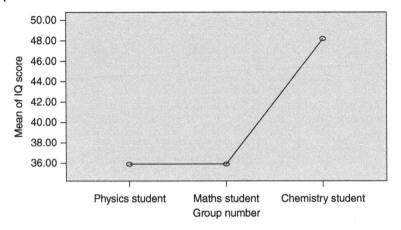

Figure 4.27 Means plot.

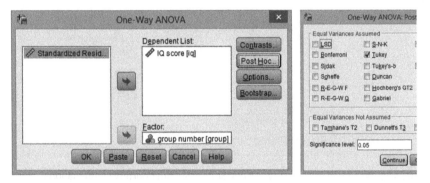

Figure 4.28 Procedure for Tukey's Post-Hoc test. Source: Reprint Courtesy of International Business Machines Corporation, © International Business Machines Corporation.

To Compute Post-Hoc Comparisons The post-hoc analysis is performed for a formal testing for group differences and to figure out which group outperforms. From the **Post-Hoc** tab of One-Way ANOVA dialog box, choose a test (**Tukey's test** is chosen here) as shown in Figure 4.28 and press Continue for the results.

Tukey's multiple comparisons post-hoc test in Table 4.30 indicates a statistically significant difference between the Chemistry students and the ones from Physics and Mathematics. The IQ score for the Chemistry students is comparatively higher.

Reporting the One-Way ANOVA Results According to the one-way ANOVA ($F(2,42)= 20.016$, p-value $= 0.000$), one can say that there was a statistically significant difference between the groups. A Tukey's post-hoc test revealed that the IQ scores for Chemistry students were statistically significantly higher as compared with the students from Physics and Mathematics. No statistically

Table 4.30 Multiple comparisons post-hoc test.

Dependent variable: IQ score

Tukey HSD

Group number (*I*)	Group number (*J*)	Mean difference (*I* − *J*)	Standard error	Sig. (*p*-value)	95% Confidence interval	
					Lower bound	Upper bound
Physics student	Maths student	−0.06667	2.25694	1.000	−5.5499	5.4165
	Chemistry student	−12.40000[a]	2.25694	0.000	−17.8832	−6.9168
Maths student	Physics student	0.06667	2.25694	1.000	−5.4165	5.5499
	Chemistry student	−12.33333[a]	2.25694	0.000	−17.8165	−6.8501
Chemistry student	Physics student	12.40000[a]	2.25694	0.000	6.9168	17.8832
	Maths student	12.33333[a]	2.25694	0.000	6.8501	17.8165

a) The mean difference is significant at the 0.05 level.

significant difference between the Physics and Mathematics students was observed.

4.4.4 What to Do if Assumption Violates?

Running a test without checking its assumptions may produce significant results (the required results); however, they might be invalid. The extent to which the assumptions violation affects results depends on the type of the test used and its sensitivity to the violation. In general, if some of the assumptions for parametric tests are not met or the data are ordinal or nominal, the alternative nonparametric tests should be used for analysis. Specifically, for one-way ANOVA, the Kruskal-Wallis nonparametric test is available in SPSS. If just the normality assumption is violated, the data should be first screened for the abnormal observations causing the nonnormality. Deletion or replacement for the detected cases in the data or using a transformation may achieve the normality, and the parametric tests can still be used.

4.4.5 What if the Assumptions in ANOVA Are Violated?

In this section, we shall investigate as to what happens if the assumptions are severely violated in applying ANOVA to solve our problem. We shall discuss this using Illustration 4.9.

Table 4.31 Learning scores of the subjects in different treatment groups.

Audiovisual	Traditional	Mixed method
40	30	5
50	34	21
52	35	23
58	36	12
52	37	24
53	32	22
54	41	31
56	39	37
42	42	23
40	48	24
59	43	57
56	50	58
57	43	89
57	42	78
58	41	66
58	40	76
59	39	79
57	42	89
59	55	85
60	80	95

Illustration 4.9 In a study, three different teaching methods were compared to see their impact on learning in a school. Three groups of students were randomly selected. These groups of students were taught the course using different teaching methods (traditional, audiovisual, and mixed method). After the training, the learning performance of the subjects in the three different groups were recorded, which is shown in Table 4.31.

We wish to test whether average performance of any treatment group is significantly better than the others. In order to test this hypothesis, let us fix $\alpha = 0.05$. We shall test the following null hypothesis:

$$H_0 : \mu_{\text{Audio\_Visual}} = \mu_{\text{Traditional}} = \mu_{\text{Mixed}}$$

against the research hypothesis that any one group mean differs.

Let us not bother about the assumptions of normality and homogeneity and apply one-way ANOVA using the SPSS software as discussed above. Here, we shall define two variables Teachning_method (independent) and

Learning_score (dependent). After making the data file as per the format given in Table 4.32, it will look like Figure 4.29.

Using the following sequence of commands in **Data View**, we shall get Figure 4.30.

Analyze → Compare Means → One Way ANOVA

After shifting variables "Learning_score" and "Teaching_method" from the left panel into "Dependent List" and "Factor" sections in the right panel and pressing the **Post Hoc**, we shall get Figure 4.31. "Tukey" option has been selected for the post-hoc test. However, readers may use any other test depending upon the situation. After pressing on **Continue**, we shall get back to the screen in Figure 4.30. Click on **Options** to get Figure 4.32 to select the option for homogeneity test and means plot.

In this screen, we shall select the options "Descriptive," "Homogeneity of variance test," and "Means plot." The homogeneity of variance test option shall provide us the output for testing the assumption of equal variance and the

Table 4.32 Data format in SPSS for one-way ANOVA.

Teaching_method	Learning_score
1	40
1	50
1	52
1	58
1	52
1	53
1	54
1	56
1	42
1	40
1	59
1	56
1	57
1	57
1	58
1	58
1	59
1	57
1	59
1	60

Table 4.32 (Continued)

Teaching_method	Learning_score
2	30
2	34
2	35
2	36
2	37
2	32
2	41
2	39
2	42
2	48
2	43
2	50
2	43
2	42
2	41
2	40
2	39
2	42
2	55
2	80
3	5
3	21
3	23
3	12
3	24
3	22
3	31
3	37
3	23
3	24
3	57
3	58
3	89
3	78
3	66
3	76
3	79
3	89
3	85
3	95

Figure 4.29 Data file in SPSS for one-way ANOVA. Source: Reprint Courtesy of International Business Machines Corporation, © International Business Machines Corporation.

	Learning_score	Teaching_method
1	40.00	1.00
2	50.00	1.00
3	52.00	1.00
4	58.00	1.00
5	52.00	1.00
6	53.00	1.00
7	54.00	1.00
8	56.00	1.00
9	42.00	1.00
10	40.00	1.00
11	59.00	1.00
12	56.00	1.00
13	57.00	1.00
14	57.00	1.00
15	58.00	1.00
16	58.00	1.00
17	59.00	1.00
18	57.00	1.00
19	59.00	1.00
20	60.00	1.00

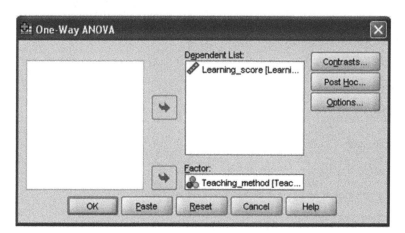

Figure 4.30 Screen showing selection of variables in one-way ANOVA. Source: Reprint Courtesy of International Business Machines Corporation, © International Business Machines Corporation.

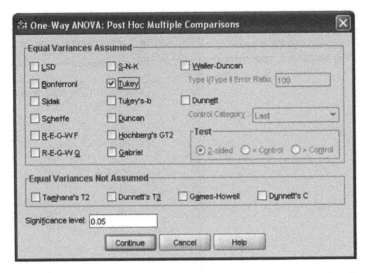

Figure 4.31 Screen showing option for Post Hoc test in one-way ANOVA. Source: Reprint Courtesy of International Business Machines Corporation, © International Business Machines Corporation.

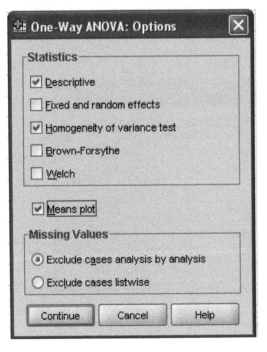

Figure 4.32 Screen showing option for homogeneity test and means plot. Source: Reprint Courtesy of International Business Machines Corporation, © International Business Machines Corporation.

Table 4.33 Test of homogeneity of variances.

Marks

Levene statistic	df1	df2	Sig. (*p*-value)
46.994	2	57	0.000

Table 4.34 ANOVA table for the scores in Maths.

Marks

	Sum of squares	df	Mean square	F	Sig. (*p*-value)
Between groups	1331.633	2	665.817	1.869	0.164
Within groups	20 301.700	57	356.170		
Total	21 633.333	59			

means plot provides the graphical representation of the means for comparison. After pressing on **Continue**, we shall get the outputs reported in Tables 4.33 and 4.34.

Table 4.34 indicates that the F-value is not significant as the significance value associated with it is 0.164, which is more than 0.05. In other words, we retain the null hypothesis and conclude that there is no evidence which indicates that any one teaching method is better than the others. In other words, variation in teaching method has no impact on learning performance.

Examining Assumptions Let us now examine two important assumptions of ANOVA, i.e. homogeneity and normality. If you look at the output in Table 4.33, it is clear that the Levene statistic is significant ($p < 0.001$). This indicates that the assumption of homogeneity has been violated. Now let us examine the assumption of normality with Shapiro test. Using the procedure discussed in Section 4.2.1, the output shown in Table 4.35 can be obtained. It can be seen from this table that the Shapiro–Wilk statistic for all the treatment groups is significant ($p < 0.05$); hence, data in each treatment group violates normality.

Since assumptions of homogeneity and normality have been violated, ANOVA should not have been used because under such circumstances, the experiment will have issue of internal validity. Under the violation of these two assumptions, we need to apply the nonparametric method for comparing the means in the three treatment groups. In this situation, we shall now test our hypothesis using the Kruskal-Wallis test.

Table 4.35 Tests of normality.

	Kolmogorov-Smirnov[a]			Shapiro–Wilk		
	Statistic	df	Sig. (*p*-value)	Statistic	df	Sig. (*p*-value)
Audiovisual	0.234	20	0.006	0.789	20	0.001
Traditional	0.279	20	0.000	0.760	20	0.000
Mixed method	0.202	20	0.031	0.889	20	0.026

a) Lilliefors significance correction.

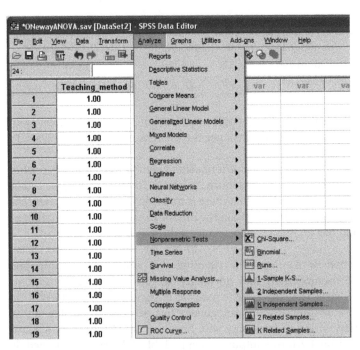

Figure 4.33 Screen showing sequence of commands in Kruskal-Wallis test. Source: Reprint Courtesy of International Business Machines Corporation, © International Business Machines Corporation.

Using the same data file shown in Figure 4.29 and following the below-mentioned sequence of commands as shown in Figure 4.33, Figure 4.34 can be obtained.

Analyze → Nonparametric Tests → k Independent Samples

By shifting the variables "Learning_score" and "Teaching_method" from the left panel to the "Test Variable List" and "Grouping Variable" sections in the right panel and defining the range of grouping variables as 1 and 3

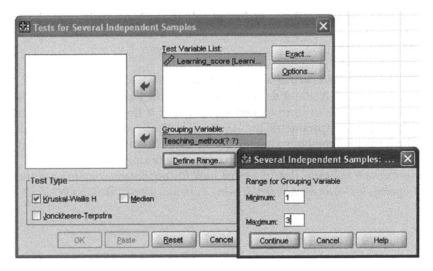

Figure 4.34 Screen showing selection of variables defining coding. Source: Reprint Courtesy of International Business Machines Corporation, © International Business Machines Corporation.

Table 4.36 Mean ranks of different groups.

	Section	N	Mean rank
Marks	Section A	20	38.55
	Section B	20	23.12
	Section C	20	29.82
	Total	60	

Table 4.37 Output of Kruskal-Wallis test.

Chi-square	7.856
df	2
Asymptotic significance	0.020[a)]

a) Significant at 5% level.

(because there are three treatment groups), we shall get the outputs as shown in Tables 4.36 and 4.37.

It can be seen from Table 4.37 that the Kruskal-Wallis statistic that has a chi-square distribution is significant ($p < 0.05$). Thus, we shall reject the null hypothesis of no difference between the three different treatment groups. In other words, all three teaching methods are not equally effective. Since there is no post-hoc comparison test in the nonparametric test, in order to know as to which treatment is most effective, comparison of the two treatment groups at a time can be carried out.

Conclusion We have seen that under the violation of homogeneity and normality assumptions, an ANOVA test indicates that all the three teaching methods are equally effective in learning performance. But if the same hypothesis is tested using the nonparametric test, i.e. Kruskal-Wallis, we concluded that all the three teaching methods are not equally effective, and any one method is superior in comparison to the others in improving the learning performance.

This exercise indicates that one must look for nonparametric option if the assumptions are severely violated in using parametric tests like ANOVA.

4.5   Correlation Analysis

Correlation is one of the statistical measures that identify the two or more variables that change together. Correlation measures the direction and magnitude or strength of the relationship between each pair of the variables. In other words, correlation is a measure of correlation or association that tests whether a relationship exists between two variables. A positive correlation shows that these variables are moving in the same direction, increasing or decreasing together, while a negative correlation means that these variables are moving in an opposite direction, one is increasing and another is decreasing.

The correlation analysis is easy and obvious, but it is very important to know which will lead you to understand your data. Readers should distinguish between the correlation as mentioned above and the causality. Causality has two components: a cause and an effect. Causality can be explained as the relationship between two events where one of them is a consequence of another. There are at least three commonly accepted conditions that must hold to be able to assume the causality: time precedence, relationship, and nonspuriousness. Thus, correlation does NOT imply causation. Other factors besides cause and effect can create an observed correlation. In other words, two variables could be correlated, but none of them causes another.

4.5.1   Karl Pearson's Coefficient of Correlation

It is known as Pearson's correlation coefficient and denoted by R. Pearson's R is the statistical measure for the association among the quantitative data. The values of the Pearson's correlation coefficient are always between -1 and $+1$. A value of $R = +1$ indicates that two variables are perfectly related in a positive linear sense. $R = -1$ means that the two variables are perfectly related in a negative linear sense, and a correlation coefficient of 0 indicates that there is no linear relationship between the two variables. The direction of the relationship is indicated by the sign of R.

$$R = \frac{\mathrm{Cov}(X, Y)}{\sqrt{\mathrm{Var}(X)\mathrm{Var}(Y)}} = \frac{\sum(X - \overline{X})(Y - \overline{Y})}{\sqrt{\sum(X - \overline{X})^2}\sqrt{\sum(Y - \overline{Y})^2}}$$

Assumptions About Data
- The two variables (X, Y) are scaled data. These variables should be neither ordinal nor nominal.
- There is no distinction between the two variables, i.e. no consideration for explanatory or response variable.
- Linear relationship exists between the two variables.
- Both variables must be normally distributed. If one or both are not, either transform the variables to near normality or use an alternative nonparametric test.

If your data violates any of the above assumptions, then use an alternative nonparametric test such as a Spearman correlation coefficient.

4.5.2 Testing Assumptions with SPSS

The first two assumptions about data (measurement scale and type) are theoretical and are important to be taken into consideration, while choosing for the parametric correlation analysis.

Illustration 4.10 The raw data given in Table 4.38 is used to illustrate testing the assumptions for correlation and regression in SPSS.

4.5.2.1 Testing for Linearity
The linearity assumption states that the relationship between each pair of correlated variables is linear. This assumption can be tested by looking at the bivariate scatter plots of the variables to be used in correlation analysis. The scatter plot uses one of the variables at x-axis and the other one at y-axis, and the observations are plotted. The resulting scatter plot showed a linear trend, i.e. the dots should be aligned in shape of a straight line. Assume that we are interested to test each bivariate correlation for the variables in Table 4.38. The scatter plot is then required for each pair. We shall first prepare the SPSS data file by defining four variables: age, mem_span (memory span), IQ, and read_ab (reading ability) as scale measures. The data file in SPSS will look like Figure 4.35. For details in preparing SPSS data file, readers can refer to Chapter 2.

The scatter plots shall be developed using the "Chart Builder." Chart Builder is an interactive chart builder window, obtained from **Graphs → Chart Builder**. It allows us to create graphs by dragging and dropping various variables/factors into the chart preview area. Below you see a Chart Builder window.

Table 4.38 Child data.

Age	Short-term memory span	IQ	Reading ability
6.70	4.40	95.00	7.20
5.90	4.00	90.00	6.00
5.50	4.10	105.00	6.00
6.20	4.80	98.00	6.60
6.40	5.00	106.00	7.00
7.30	5.50	100.00	7.20
5.70	3.60	88.00	5.30
6.15	5.00	95.00	6.40
7.50	5.40	96.00	6.60
6.90	5.00	104.00	7.30
4.10	3.90	108.00	5.00
5.50	4.20	90.00	5.80
6.90	4.50	91.00	6.60
7.20	5.00	92.00	6.80
4.00	4.20	101.00	5.60
7.30	5.50	100.00	7.20
5.90	4.00	90.00	6.00
5.50	4.20	90.00	5.80
4.00	4.20	101.00	5.60
5.90	4.00	90.00	6.00

Follow the given sequence to proceed to the Chart Builder as shown in Figure 4.36. After filling the desired entries, we shall get the scatter plots as shown in Figure 4.37.

By looking at the scatter plot, one can see that a perfect linearity is not found because all the dots are not perfectly aligned around the straight line. However, the linearity is exhibited for some pairs of variables up to some extent. The correlation between the pairs of linearly related variables will be determined by computing Karl Pearson's coefficient using SPSS.

The Karl Pearson's coefficient has been illustrated using the child data set (Table 4.38). Usually, the significance value for both coefficients coincide with each other; however, the Pearson's parametric correlation is appropriate for normally distributed data. The Spearman correlation is nonparametric and may be applied on categorical variables. From the SPSS menu, follow the sequence as in Figure 4.38: **Analyze → Correlate → Bivariate**.

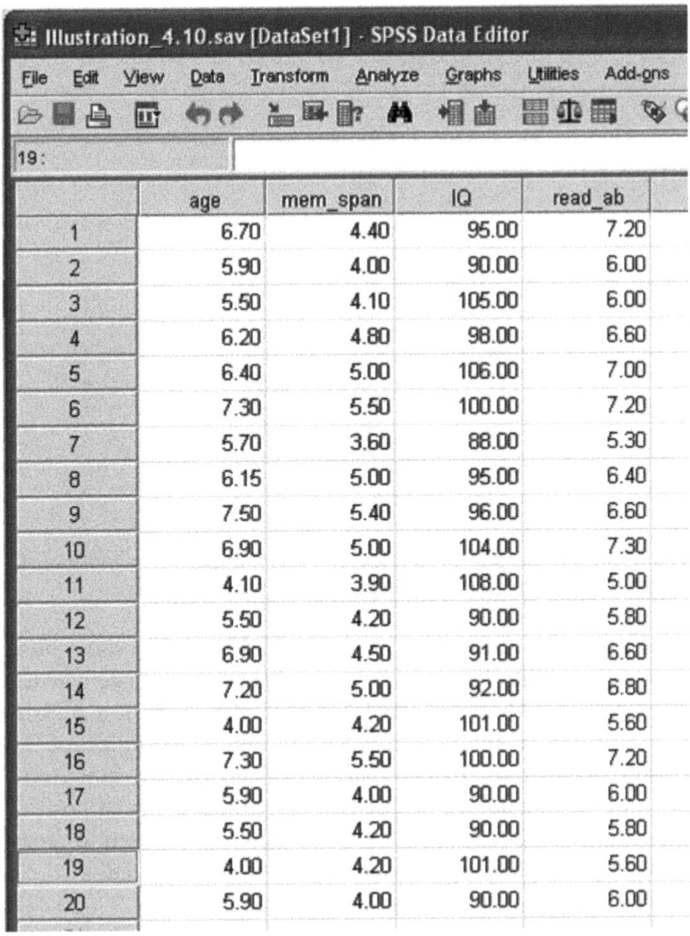

	age	mem_span	IQ	read_ab
1	6.70	4.40	95.00	7.20
2	5.90	4.00	90.00	6.00
3	5.50	4.10	105.00	6.00
4	6.20	4.80	98.00	6.60
5	6.40	5.00	106.00	7.00
6	7.30	5.50	100.00	7.20
7	5.70	3.60	88.00	5.30
8	6.15	5.00	95.00	6.40
9	7.50	5.40	96.00	6.60
10	6.90	5.00	104.00	7.30
11	4.10	3.90	108.00	5.00
12	5.50	4.20	90.00	5.80
13	6.90	4.50	91.00	6.60
14	7.20	5.00	92.00	6.80
15	4.00	4.20	101.00	5.60
16	7.30	5.50	100.00	7.20
17	5.90	4.00	90.00	6.00
18	5.50	4.20	90.00	5.80
19	4.00	4.20	101.00	5.60
20	5.90	4.00	90.00	6.00

The SPSS Data Editor shows: Illustration_4.10.sav [DataSet1] - SPSS Data Editor, with menu File Edit View Data Transform Analyze Graphs Utilities Add-ons, and cell reference 19:

Figure 4.35 Data file in SPSS for correlation and regression analysis. Source: Reprint Courtesy of International Business Machines Corporation, © International Business Machines Corporation.

In the **Bivariate Correlations** window shown in Figure 4.39, choose the variables to be tested for bivariate correlations from the "**Variables**" section. Since the linearity assumption is not satisfied for some pairs of variables, the Spearman's nonparametric correlation is also selected in addition to the parametric Pearson's correlation. Press **Continue** for the output in Tables 4.39 and 4.40.

a) *Interpretation*: The correlations with a p-value (p) less than 0.05 are said to be significant. An absolute value closer to 1 is said to be a strong correlation. The statistically significant strong linear correlations have been observed

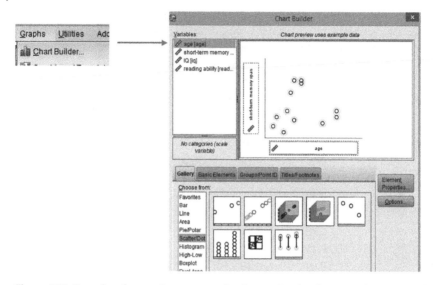

Figure 4.36 Procedure for creating a scatter plot. Source: Reprint Courtesy of International Business Machines Corporation, © International Business Machines Corporation.

between the pairs age $<->$ short-term memory, reading ability; short-term memory $<->$ reading ability, based on Spearman's correlation coefficient.

b) *Reporting*: The final lines to be reported are presented here in a statistical way: A Pearson product–moment correlation was run to determine the linear relationship between the pairs of variables from age, IQ, short-term memory span, and reading ability. A statistically significant strong positive linear correlation is observed between the age and short-term memory span ($r = 0.723$, $n = 20$, $p = 0.000$), age and reading ability ($r = 0.846$, $n = 20$, $p = 0.000$), and short-term memory span and reading ability ($r = 0.821$, $n = 20$, $p = 0.000$).

In this context, there is a very important measure, which is frequently utilized for measuring the goodness of fit, which is the "Coefficient of Determination."

4.5.3 Coefficient of Determination

Coefficient of determination (R^2) is the most suitable and clear way of understanding the value of correlation coefficient using the square of linear correlation coefficient. Suppose $R = 0.9$, then $R^2 = 0.81$; this would interpret as 81% of the variation in the dependent variable has been explained by the independent variable while 19% is due to other variables that are not in the model or the study.

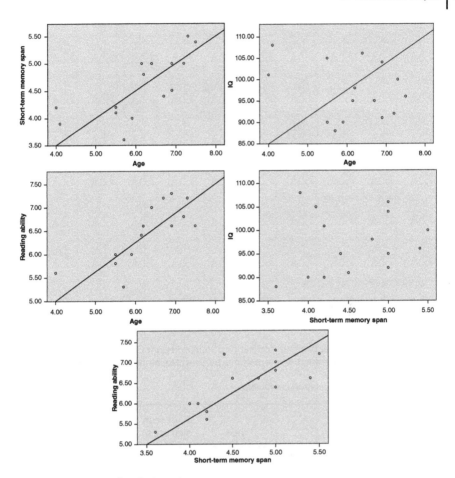

Figure 4.37 Scatter plots for linearity.

Figure 4.38 Path for bivariate correlations. Source: Reprint Courtesy of International Business Machines Corporation, © International Business Machines Corporation.

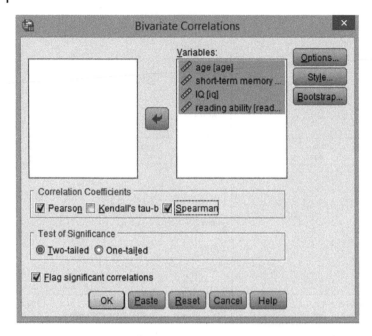

Figure 4.39 Procedure for bivariate correlations. Source: Reprint Courtesy of International Business Machines Corporation, © International Business Machines Corporation.

Table 4.39 Pearson's correlation coefficient.

		Age	Short-term memory span	IQ	Reading ability
Age	Pearson correlation	1	0.723[a]	−0.198	0.846[a]
	Sig. (two-tailed)		0.000	0.402	0.000
	N	20	20	20	20
Short-term memory span	Pearson correlation	0.723[a]	1	0.302	0.821[a]
	Sig. (two-tailed)	0.000		0.195	0.000
	N	20	20	20	20
IQ	Pearson correlation	−0.198	0.302	1	0.150
	Sig. (two-tailed)	0.402	0.195		0.527
	N	20	20	20	20
Reading ability	Pearson correlation	0.846[a]	0.821[a]	0.150	1
	Sig. (two-tailed)	0.000	0.000	0.527	
	N	20	20	20	20

a) Correlation is significant at the 0.01 level (two-tailed).

Table 4.40 Spearman's correlation coefficient.

			Age	Short-term memory span	IQ	Reading ability
Spearman's rho	Age	Correlation coefficient	1.000	0.777[a]	0.004	0.881[a]
		Sig. (two-tailed)		0.000	0.987	0.000
		N	20	20	20	20
	Short-term memory span	Correlation coefficient	0.777[a]	1.000	0.340	0.803[a]
		Sig. (two-tailed)	0.000		0.143	0.000
		N	20	20	20	20
	IQ	Correlation coefficient	0.004	0.340	1.000	0.214
		Sig. (two-tailed)	0.987	0.143		0.365
		N	20	20	20	20
	Reading ability	Correlation coefficient	0.881[a]	0.803[a]	0.214	1.000
		Sig. (two-tailed)	0.000	0.000	0.365	
		N	20	20	20	20

a) Correlation is significant at the 0.01 level (two-tailed).

The maximum value of R^2 is 1 because it is possible to explain all of the variation in Y but it is not possible to explain more than all of it: Coefficient of determination = Explained variation/Total variation.

4.6 Regression Analysis

In this section, we will review the basics of the simple linear regression model and how to implement and test it using the SPSS. Regression model is a statistical technique for assessing the relationship between dependent or response variable and one or more independent variables. The relationship between two variables is characterized by how they vary together. One common problem in research is to determine whether a functional relationship exists between two variables and, if so, to characterize and quantify this relationship. For example, we may desire to study the relationship between the amount of money that a company spends on advertising its product and the amount of sales of the product. Hopefully, the more we tell people about the product, the more they want to buy it. Regression models can be used to quantify and evaluate the relationship between these variables.

We label the two variables as X and Y, where the X variable is often called the "independent" or "regressor" variable. The Y variable is often called the "dependent" or "response" variable. These two variables are distinguished by the fact that the X variable is the variable that can be controlled by the researcher whereas the Y variable reacts or changes due to the changes in the X variable. In our example above, the X variable is the amount of money spent on advertising and the Y variable is the amount of sales of our product. It is desired to express Y as some function of X, say $Y = f(X)$, where f is some arbitrary function; here, we are assuming it to be a linear function.

A common statistical model appropriate for the above situation is

$$Y = f(X) + \varepsilon,$$

where ε represents the random error term, usually assumed to have a mean of zero and a variance of σ^2. Thus, according to our model, the expected value of Y, at a fixed value of X, denoted by $E(Y|X)$, equals $f(X)$ and the variance of Y, at the fixed value of X, denoted by $\text{Var}(Y|X)$, is σ^2. It is noted that while the mean of Y is a function of X and varies according to the function $f(X)$, the variance is constant across all values of X.

It is certainly possible that several other regressor/independent variables, in addition to the variable X, are acting together to drive the response/dependent variable Y. In this case, our model might become

$$Y = f(X_1, X_2, \dots, X_k) + \varepsilon,$$

where $X_1, X_2, \dots, X_k$ represents k regressors or independent variables. While it is possible to deal with the above statistical model using the single variable X and the arbitrary function $f(X)$, we will first address the situation where the function $f(X)$ is defined as the linear function

$$f(X) = \beta_0 + \beta_1 X.$$

Here, $f(X)$ is recognized as the formula for the straight line in X, with the coefficients β_0 and β_1 representing the intercept and the slope of this line, respectively. Assuming the variable Y as a response and X as an explanatory variable, the regression analysis can be used to measure the direction (positive and negative) and the rate of change in Y as X changes (slope). Regression model is appropriate for predicting the value of Y, given X rather than just a measure of association or the strength of relationship between them.

4.6.1 Simple Linear Regression

Simple linear regression model, a special case of multiple linear regression when there is only one independent variable, is a statistical method that allows researchers to summarize and study relationships between two continuous (quantitative/scale) variables. In a cause and effect relationship, the independent variable is the cause and the dependent variable is the effect. Least squares linear regression is a method for predicting the value of response, outcome, or dependent variable Y, based on the value of an independent, predictor, or explanatory variable X.

Our statistical model now becomes

$$Y = f(X) + \varepsilon = \beta_0 + \beta_1 X + \varepsilon,$$

commonly referred to as the "simple linear regression" model. The "simple" refers to the use of one regressor in the model (as opposed to using k regressors as in the "multiple linear regression") and "linear" refers to the "linearity in the regression coefficients." By this we mean that each coefficient β_0 or β_1 enters the regression model raised to power 1.

The researchers/practitioners should interpret the regression model as follows:

- β_0 is the "intercept" or "constant," which is the point at which the regression intercepts y-axis. Intercept provides a measure about the mean of dependent variable when independent variable(s) are zero. If slope(s) are not zero, then intercept is equal to the mean of dependent variable minus slope × mean of independent variable.
- β_1 is the "slope" or "regression coefficient," which is the change in the dependent variable as we change the independent variable by 1 unit. $\beta_1 = 0$ means that the independent variable does not have any influence on the dependent variable.

For estimating the regression model and to obtain

$$\hat{y}_i = \hat{\beta}_0 + \hat{\beta}_1 x_i, \quad i = 1, 2, \dots, n.$$

The method of ordinary least squares (OLSs) is usually used to accomplish this task. The main idea for this method is to choose the values of the coefficients so that the quantity

$$\sum_{i=1}^{n} (y_i - \hat{y}_i)^2$$

is minimized. Here, $\hat{y}_i$ denotes the model's fitted value or "estimated value" to the response at x_i. That is, $\hat{y}_i = \hat{\beta}_0 + \hat{\beta}_1 x_i$, where $\hat{\beta}_0$ and $\hat{\beta}_1$ denote the OLS estimates of the coefficients. The quantity $y_i - \hat{y}_i$, denoted by $\hat{\varepsilon}_i$, is called the residual for the ith observation. Obviously, $\hat{\varepsilon}_i$ indicates the degree to which the model is able to fit y_i, with small values indicating a "good" fit and large values indicating a "poor" fit. The minimized quantity, $\sum_{i=1}^{n} (y_i - \hat{y}_i)^2 = \sum_{i=1}^{n} \hat{\varepsilon}_i^2$, represents the model's total inability to fit the data across all n observations.

4.6.2 Assumptions in Linear Regression Analysis

There are several assumptions that need to be validated before performing the regression model. The regression model is based on the following assumptions:

1) *Linearity*: The relationship between Y and X is linear relationship (straight-line relationship).
2) Errors (ε) or residuals (estimated errors) are normally distributed.
3) *Homoscedasticity*: The variance of the residuals is equal for all X.
4) There is no autocorrelation ($\text{cov}(\varepsilon_i, \varepsilon_j) = 0$, i ≠ j).
5) The expected value of the error (ε) term is zero ($E(\varepsilon) = 0$).
6) There is no measurement error on X (impractical assumption).
7) In the multiple regression model, if there are more than one independent variables in the equation, then the X variables should not be perfectly correlated to each other.

4.6.2.1 Testing Assumptions with SPSS

Assumption 1: The readers know that the simple linear regression is performed after $n(x,y)$ pairs of observations are taken on n independent experimental units. Our data is represented as the n pairs (x_i, y_i), for $i = 1, ..., n$. The first step in the analysis to validate the above assumptions is graphical, where these n pairs of points are plotted on the usual x-y coordinate system. The result is a scatter of n points, often called a "scatter plot" or "scatter diagram." Visual inspection of this scatter plot helps to affirm the use of the simple linear regression model and the validity of the resulting statistical analysis. According to the simple linear regression model, the data should behave as though they are random observations about the true line represented by $Y = \beta_0 + \beta_1 X$. Thus, we would hope that our scatter plot also exhibits this tendency. If so, we continue with the analysis that usually includes obtaining estimates of the model coefficients, performing inference on one or both of the coefficients, and evaluating quantitative measures expressing the quality of the model.

If the scatter plot indicates that the data does not conform to the straight-line model, then there are a number of options available to the researcher. One option, if appropriate, would involve transforming either the Y or the X variable, so that the transformed data more closely conforms to the simple linear regression model. In this case, the analysis proceeds as before but on the transformed data. A second option is to use another model such as a nonlinear model or use a multiple linear regression model or consider an error distribution other than the normal such as a heavy-tailed distribution, a binomial distribution, or a Poisson distribution.

Assumption 2: Errors (ε) or residuals (estimated errors) are normally distributed at every value of the dependent variable and distributed symmetrically. If there are any violations, such as skewness (nonsymmetrically, one tail longer than the other), kurtosis (too flat or too peaked), and existence of any outliers (individual cases that are far from the distribution), the users should examine this assumption in the univariate case using the histograms, box plots, and P–P plots or Q–Q plots for the residuals or response variable.

Assumption 3: Homoscedasticity – The variance of the residuals is equal for every set of values when the independent variable (X) is equal. This assumption could be investigated by drawing a scatter plot, which works only with one independent variable, for the predicted values against residuals or standardized residuals or deleted residuals or standardized deleted residuals. The user should have expected a completely random pattern for the homoscedasticity. Otherwise, this assumption will be violated, which means there is a "heteroscedasticity."

Assumption 4: The expected autocorrelation between residuals, for any two cases, is 0 $(\text{cov}(\varepsilon_i, \varepsilon_j) = 0)$. All observations should be independent of one another. In other words, knowing the value of one observation should not tell you anything about the value of other observations. To detect this problem, residual plots and Durbin Watson (DW) test should be utilized.

Assumption 5: The expected value of the error (ε) term is zero $(E(\varepsilon) = 0)$. This assumption mainly affects the constant term in the regression model. So, if the mean of the error term deviates from zero, the constant soaks it up.

Assumption 6: All independent variables are uncorrelated with the error term. In other words, there is no measurement error on X (impractical assumption). By definition, the residuals are uncorrelated with the independent variables (try it and see, if you like). There are several applications where this assumption is unrealistic such as demand increases supply, supply increases wages, and higher wages increase demand. Consequently, the OLS estimates will be (badly) biased estimators in these cases. Hence, the user needs different estimation procedure like two-stage least squares.

Assumption 7: None of the independent variables is a perfect linear function of other independent variables. In the multiple regression model, there are more than one independent variables in the equation and the X variables should not be perfectly correlated to each other. In other words, there should not be any multicollinearity problem or no linear relationships among the independent variables, i.e. $X_1 = X_2 + 5$ while $X_1 = \log(X_2)$ is not linear relationship and hence does not cause a multicollinearity problem. To detect this problem, users should look over the variance inflation factor (VIF). Also, there are some symptoms for having high degree of multicollinearity such as if the overall model is highly significance while most of the independent variables are not. Some of signs for the independent variables are contradicting the theory. In the following, the users will see how to perform the assumptions diagnostics using the SPSS.

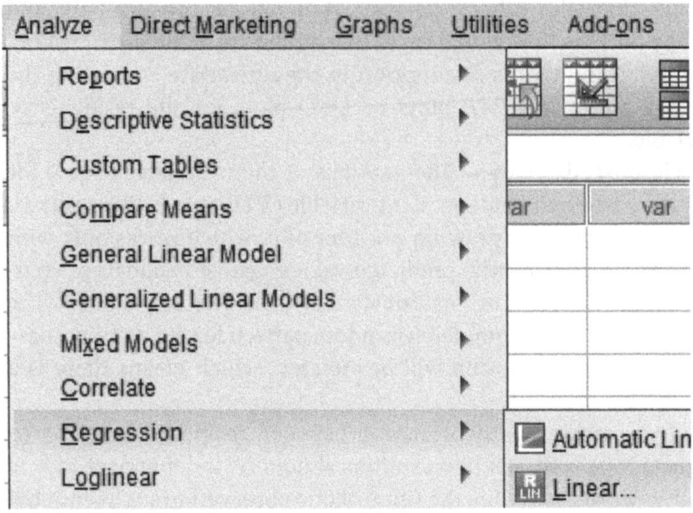

Figure 4.40 Path for linear regression analysis. Source: Reprint Courtesy of International Business Machines Corporation, © International Business Machines Corporation.

Illustration 4.11 Multiple Linear Regression: Consider the same SPSS data file 4.10 used in illustration 4.10 shown in Figure 4.35 used for correlations. Assume that we want to fit a regression model, where the dependent variable is the reading ability of a child that depends on his or her age and short-term memory span (independent variables). The linearity assumption is already tested using scatter plots. The normality of errors can be tested on the standardized residuals saved from the regression analysis. For testing the assumptions on residuals, start with the regression analysis by choosing the following analysis options from the SPSS menu: **Analyze → Regression → Linear** (as in Figure 4.40).

Select the "Reading Ability" as dependent and "Age" and "Short-term Memory Span" as independent variables (as in Figure 4.41).

By clicking on the **Statistics** tab, you will get Figure 4.42. Then choose "Estimates," "Model fit," "Collinearity Diagnostics" (for testing multicollinearity), and "Durbin-Watson" (for testing autocorrelation) and click on **Continue**.

By clicking on the plots tab in Figure 4.41, you will get a window (Figure 4.43). Then choose the "ZPRED" in variable "X" and "ZRESID" in variable "Y" for having a residual plot. Also check for histogram and normal probability plot for testing the normality of errors. Click on **Continue** for the analysis results (Figures 4.44 and 4.45 and Tables 4.41–4.45).

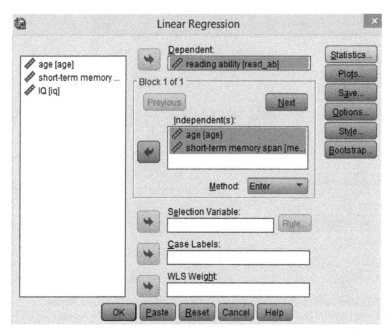

Figure 4.41 Procedure for Linear Regression analysis. Source: Reprint Courtesy of International Business Machines Corporation, © International Business Machines Corporation.

Figure 4.42 Procedure for testing autocorrelation and multicollinearity. Source: Reprint Courtesy of International Business Machines Corporation, © International Business Machines Corporation.

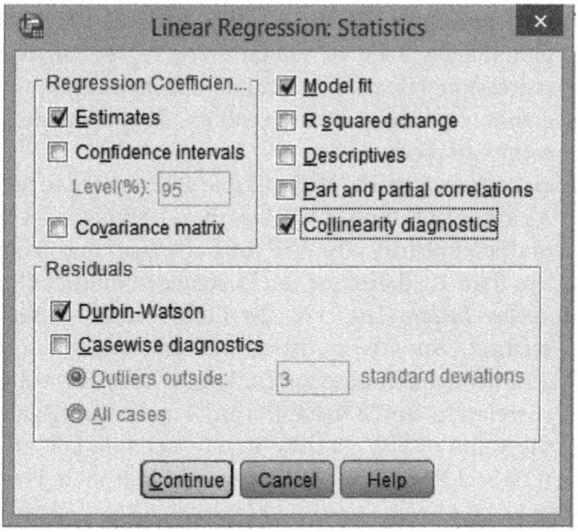

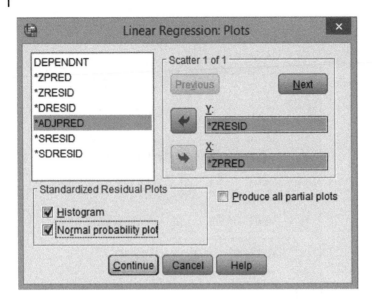

Figure 4.43 Procedure for testing the normality of errors assumption. Source: Reprint Courtesy of International Business Machines Corporation, © International Business Machines Corporation.

Normality of errors: The histogram and the normal probability plot shows that the errors are normally distributed, satisfying the stated assumption for regression analysis.

Homoscedasticity: The homoscedasticity can be visually tested by the scatter plot (Figure 4.45) of standardized regression residuals and the predicted regression values. The scatter plot in the output shows no pattern in the errors, i.e. the errors are random. This confirms the assumption for equal error variances.

Multicollinearity: The VIF, in Table 4.41, is used to test for the multicollinearity. As a rule of thumb, a VIF less than 10 is considered as an acceptable level of multicollinearity. The VIF for both "age" and "short-term memory span" is less than 10; therefore, the existence of multicollinearity can be ignored.

Autocorrelation: The DW test (Table 4.42) is used for testing the autocorrelation. The DW statistic could take any value from 0 to 4. A value near 2 indicates non-autocorrelation, a value toward 0 indicates positive autocorrelation, and a value toward 4 indicates negative autocorrelation. Since the value of DW statistic in our example falls under the acceptable range (DW = 1.912–2), it can be assumed that there is no autocorrelation.

Regression analysis results and interpretation: The ANOVA Table 4.43 presents the goodness of fit for the overall regression model. The overall regression model is significant at 5% level ($F(2,17) = 35.568$, p-value = 0.000 < 0.05).

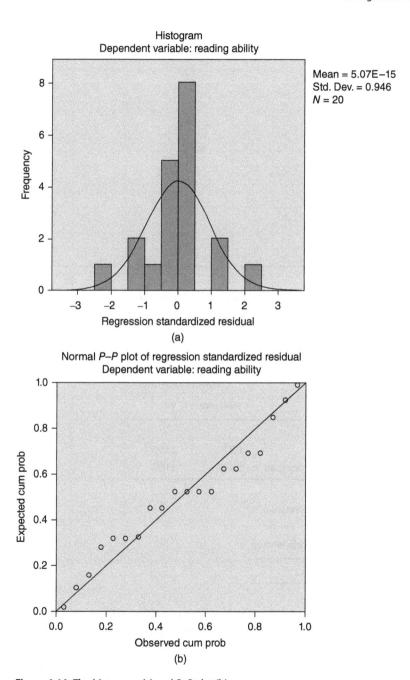

Figure 4.44 The histogram (a) and *P–P* plot (b).

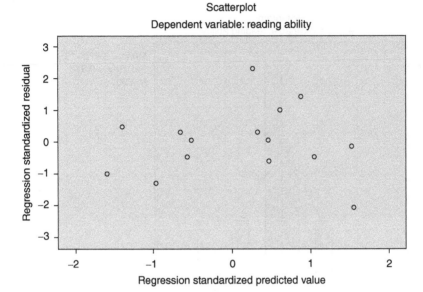

Figure 4.45 Scatter plot for the standardized residual.

Table 4.41 Testing for multicollinearity.

Model		Collinearity statistics	
		Tolerance	VIF
1	(Constant)		
	Age	0.477	2.095
	Short-term memory span	0.477	2.095

Table 4.42 Testing for autocorrelation.

Model	Durbin-Watson
1	1.912

A higher value of the coefficient of determination (R^2) in Table 4.44 shows a strong combined impact of independent variables "age" and "short-term memory span" on the dependent variable "reading ability." It can be stated that the short-term memory span and age of the child together explain 80.7% of the variation in child's reading ability while 19.3% is due to other variables that are not in the model.

Table 4.43 The ANOVA table: testing for overall model fitness.[a]

Model		Sum of squares	df	Mean square	F	Sig. (p-value)
1	Regression	7.248	2	3.624	35.568	0.000[a]
	Residual	1.732	17	0.102		
	Total	8.980	19			

a) *Predictors*: (constant), short-term memory span, age.
b) *Dependent variable*: reading ability.

Table 4.44 Model summary.[b]

Model	R	R²	Adjusted R²	Standard error of the estimate
1	0.898[a]	0.807	0.784	0.31920

a) *Predictors*: (constant), short-term memory span, age.
b) *Dependent variable*: reading ability.

Table 4.45 Regression coefficients.[a]

Model		Unstandardized coefficients B	Standard error	Standardized coefficients β	t	Sig. (p-value)
1	(Constant)	1.897	0.578		3.281	0.004
	Age	0.339	0.099	0.530	3.435	0.003
	Short-term memory span	0.521	0.184	0.438	2.839	0.011

a) *Dependent variable*: reading ability.

It can be seen from Table 4.45 that the constant (model intercept), age, and short-term memory span have a statistically significant impact on the reading ability of a child at 5% level of significance. Based on the unstandardized coefficients, it can be concluded that on average a unit (year) increase in child's age can cause an increase in reading ability by 0.339 units. A unit increase in the short-term memory span can increase reading ability by 0.521 units. The standardized coefficients indicate the same but in terms of standard deviations instead of units and are used to find the relative importance of the variables in the model.

Reporting: A multiple linear regression was calculated to predict the child's reading ability based on their age and short-term memory span. A significant regression equation was found ($F(2,17) = 35.568$, p-value $= 0.000 < 0.05$) with an R^2 of 80.7%.

Exercises

Multiple-Choice Questions

Note: Choose the best answer for each question. More than one answer may be correct.

1. Statistical analyses can be divided into two well-known categories named as
   a. Frequencies (mean, median, mode)
   b. Inferential and descriptive statistics
   c. Graphs and charts
   d. Qualitative and quantitative

2. The set of assumptions make a distinction between two different procedures for inferential statistics called
   a. Nonparametric and parametric techniques
   b. Hypothesis testing
   c. Graphical techniques
   d. Confidence interval

3. Choice between parametric and nonparametric tests depends on
   a. Assumptions regarding data
   b. Number of variables
   c. Measurement scale of data
   d. Objective of the study

4. Creativity of a person can best be measured on
   a. Interval scale
   b. Ratio scale
   c. Numerical scale
   d. Ordinal scale

5. Common assumptions in parametric tests are
   a. Homogeneity of variances
   b. Randomness
   c. Nonlinearity
   d. Normality

6. In SPSS, the normality assumption can be tested using
   a. Levene's test
   b. Q–Q plot
   c. Box plot
   d. Shapiro–Wilk test

7. What to do if normality assumption is violated?
   a. Stop the analysis
   b. Ignore and get the results
   c. Apply transformations
   d. Use parametric tests

8. In SPSS, the homogeneity of variances can be tested using
   a. Levene's test
   b. Kolmogorov–Smirnov test
   c. Independent-samples t-test
   d. Box's M test

9. In SPSS, the linearity assumption can be tested using
   a. Runs test
   b. Scatter plot
   c. Significance of correlation
   d. Significance of regression coefficients

10. Commonly used statistics for hypothesis testing for independence is
    a. Runs test
    b. F-test
    c. t-Test
    d. z-Test

11. Both the z-test and t-test are appropriate for
    a. Hypothesis testing
    b. Small sample size
    c. Comparing the group means
    d. Comparing the group variances

12. The F-test is used for
    a. Hypothesis testing
    b. Goodness-of-fit test for regression
    c. Comparing the group means
    d. Comparing the group variances

13. In each case for the comparison of group means, which option(s) suggest an appropriate test?
a. 1 Group: z-test or t-test
b. 2 Groups: t-test or F-test
c. More than 2 groups: t-test
d. More than 2 groups: F-test

14. The analysis of variance (ANOVA) is used for
a. Comparing group variances
b. Comparing group means
c. Testing within group variations
d. Testing between group variations

15. While using Shapiro–Wilk test for testing normality, assumption holds if we
a. Reject the null hypothesis (p-value $< \alpha$)
b. Fail to reject the null hypothesis (p-value $> \alpha$)
c. Have bimodal distribution
d. Have skewed distribution

16. The assumption for equality of variances holds if
a. We reject the null hypothesis (p-value $< \alpha$)
b. There is large variability between groups
c. There is large variability due to treatments
d. We fail to reject the null hypothesis (p-value $> \alpha$)

17. In ANOVA, the *post-hoc* comparisons are to be carried out when
a. The null hypothesis for equality of means is rejected (p-value $< \alpha$).
b. We fail to reject the null hypothesis for equality of means (p-value $> \alpha$), concluding that at least one mean is different.
c. The treatments have no effect.
d. All options are correct.

18. The Pearson bivariate correlation coefficient could be calculated if
a. One of the variables is dichotomous
b. Both the variables are metric
c. One of the variables is continuous and the other one is dichotomous
d. All the three variables are dichotomous

19. Which one of the following is required for doing hypothesis testing and confidence interval for linear regression model?
a. Normality
b. Linearity

c. Heteroscedasticity
d. Autocorrelation and multicollinearity

20. The regression analysis is said to be "multiple linear regression analysis" when we have
    a. One dependent and one independent variable
    b. One dependent and two or more independent variables
    c. Two dependent and one or more independent variables
    d. In all of the above cases

Short-Answer Questions

1. Discuss the importance of assumptions in research. What are the common assumptions in using parametric tests and which test will you use to test each of them?

2. Explain different graphical methods of testing normality.

3. If normality assumptions are violated, what remedial measures you will adopt? In that situation, which statistics will you report in your research report and why?

4. What is an outlier and how does it affect the findings? Discuss it by means of example.

5. What is Run and how it is calculated? Explain the procedure in applying Runs test for testing randomness.

6. What do you mean by randomization and how is it done in designing the hypothesis testing experiments? What are the benefits of randomization in research study?

Answers

Multiple-Choice Questions

1. b

2. a

3. c

4. d

5. b, d

6. d

7. c

8. a

9. b

10. a

11. c

12. d

13. a, d

14. b, d

15. b

16. d

17. a

18. b

19. a

20. b

5

Assumptions in Nonparametric Tests

5.1 Introduction

Statistical inference is used for generalizing results from the sample to the population. Most of the statistical tests depend on two main assumptions: randomness and normality. However, in some situations, these assumptions are violated and using the parametric tests will not be appropriate. Therefore, statisticians introduced the nonparametric techniques to deal with such cases, where the above assumptions aren't met. The nonparametric statistics is known to be distribution free.

The nonparametric statistics is appropriate when the population's parameters do not satisfy certain criteria, i.e. not measurable, such as in the categorical data, where the mean and variance are not defined. In other words, if there is a small sample size and non-normal distribution, the nonparametric techniques have to be utilized for making an inference about the population's parameters. The readers should know that the nonparametric tests are less precise when compared with the parametric tests because there is less known information about the population.

In this chapter, the readers will get familiar with the required nonparametric assumptions, different nonparametric tests, how to perform those using IBM SPSS®[1] Statistics software (SPSS), and what should be done if there is any violation for these assumptions.

5.2 Common Assumptions in Nonparametric Tests

The nonparametric tests rely on fewer assumptions about the sample data used for drawing inferences about the population, as compared with the parametric tests. However, there are certain assumptions attached to the nonparametric tests too. The common assumptions in nonparametric tests are Randomness

---

1 SPSS Inc. was acquired by IBM in October 2009.

Testing Statistical Assumptions in Research, First Edition. J. P. Verma and Abdel-Salam G. Abdel-Salam.
© 2019 John Wiley & Sons, Inc. Published 2019 by John Wiley & Sons, Inc.
Companion Website: www.wiley.com/go/Verma/Testing_Statistical_Assumptions_Research

and Independence. These two assumptions are related to the sample but not the population, as the nonparametric tests make no assumptions about the underlying population.

5.2.1 Randomness

Randomness is one of the essential assumptions to be met to have reliable and valid statistical results. Randomness refers to drawing a random sample from the population, without selection bias. In experimental design, randomness refers to the random assignment of individuals or subjects to the treatment groups, where the only difference between groups is due to the treatment. Consider an example of public polling about a product, and all 20 respondents turn out to be males. This will distort the assumed randomness of the sample because of the same gender.

5.2.2 Independence

The second common assumption is *independence*, referring to the statistically independent sampling trials. In other words, the observations between groups should be independent; the groups must not present the responses from same individuals.

5.2.2.1 Testing Assumptions Using SPSS

The subsequent section covers the discussion over testing common assumptions for nonparametric tests using SPSS. The SPSS sample data file *"Illustration 5.1.sav"* is used for the application of runs test. The example data is originally a survey data, containing the information about general beliefs of people and their demographics (living in the United States) subject to their country of origin. To start with testing, the assumptions open the sample data file *"Illustration 5.1.sav"* from the *Program Files* folder, where the SPSS software is installed.

Tests of randomness are important addition to the statistical theory because the requirement for random sample is essential in almost all of the statistical tests. The techniques used for testing randomness are called the *runs up and down test* and the *rank von Neumann test*.

In SPSS, the "Runs test" is used for testing if the order of occurrence of two values of a variable is random. The readers could conclude that the sample is not random if it has too many or too few runs.

Illustration 5.1 Consider the data set given in SPSS file *"Illustration_5.1.sav"* (*available on the companion website of this book*); the data set have been used from the site gss.norc.org with the permission of GSS. Since the data set is based on public opinions, the hypothesis for randomness can be tested based on the *Gender* and *Age* of the respondents. The null hypothesis for testing the randomness can be stated as "the data follows a random sequence." In other words, the null hypothesis here assumes that the data comes from a random sample, based on *Gender* and/or *Age*.

5.2.2.2 Runs Test for Randomness Using SPSS

To perform the nonparametric Runs test in SPSS, follow the following steps (Figure 5.1):

Analyze → Nonparametric Tests → Legacy Dialogs → Runs

- The Runs test dialog box will appear (as in Figure 5.2). Choose the variables "age" and "sex" in "Test Variable List" section to check the randomness of

Figure 5.1 Path for the Runs test. Source: Reprint Courtesy of International Business Machines Corporation, © International Business Machines Corporation.

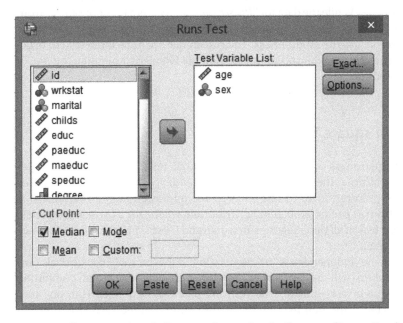

Figure 5.2 Choosing options for Runs test. Source: Reprint Courtesy of International Business Machines Corporation, © International Business Machines Corporation.

Table 5.1 Runs test for the Gender and Age of respondents.

	Age of respondent	Gender
Test value[a]	42	2
Cases < test value	1357	1232
Cases ≥ test value	1471	1600
Total cases	2828	2832
Number of runs	1302	1417
Z	−4.171	0.914
Asymptotic Significance (two-tailed)	0.000	0.361

a) Median.

sample and choose option for "Cut Point" to calculate the runs. No assumption regarding the population distribution is made here; therefore, we choose Median and click **OK**. Then the output will be as in Table 5.1.

- *Interpretation*: The null hypothesis for Runs test assumes that the "data is random." Based on p-values from Table 5.1, we may conclude that the observations in variable Age ($p < 0.05$) do not conform to the randomness assumption for Age. While the data for "*Illustration 5.1.sav*" is a random sample from the population of males and females, the age is not the variable of interest with respect to the randomness. The researcher might be interested to make inferences about public opinion based on the Gender, rather than the Age of the respondents.

5.3 Chi-square Tests

The chi-square test is one of the nonparametric tests for testing three types of statistical tests: the goodness of fit, independence, and homogeneity. This goodness-of-fit test is for testing whether the interested random variable(s) is derived from a specific probability distribution, where, for a certain categorical variable, it tests if all the categories are equal and there is no statistical difference between them.

Testing the independence or association between two categorical random variables will be performed by doing the crosstabs and computing the chi-square test statistic. The test compares the expected (the one that you expected if the null hypothesis is true) and observed (actual number in each category in the data) frequencies in each category for testing if all the categories contain same proportion of values or the user-specified proportion of values. The null hypothesis states that the observed frequencies are equal to the expected frequencies.

To test for homogeneity, one would draw independent samples of females and males and would classify them according to the teaching track (arts or sciences). Therefore, each row is a separate binomial distribution, and the researchers want to see if these binomial distributions are identical. One should know that the test of independence and test of homogeneity have the same calculated chi-square and same degrees of freedom ((number of rows $-$ 1) $\times$ (number of columns $-$ 1)). Hence, the two tests are equivalent. However, they have different hypotheses and different interpretations.

The general chi-square statistic's formula is given by

$$\chi^2 = \sum \frac{(\text{Observed frequency} - \text{Expected frequency})^2}{\text{Expected frequency}}.$$

5.3.1 Goodness-of-Fit Test

This goodness-of-fit test categorizes a variable and tabulates it for computing a chi-square statistic. The test compares the observed and estimated/expected frequencies in each category to test whether all categories have the same or user-specified proportion of values. The null hypothesis states that the observed frequencies are equal to the expected frequencies; in other words, for a certain categorical variable, it tests that all the categories are equal and there is no statistical significance difference between them.

5.3.1.1 Assumptions About Data
- The numerical categorical variables are measured either at ordinal or nominal levels and can be used for the goodness-of-fit chi-square test.
- There is no requirement for the shape of underlying distribution to conduct the goodness-of-fit test, but the data should be based on a random sample.
- The data in the cells should be frequencies or counts and not the percentages or some other transformation of data.
- The expected frequencies for each category should appear only once in the data, i.e. the study groups must be independent. In other words, the chi-square test is not appropriate for *paired* samples.
- For two categories, the expected frequencies at each cell should not be less than 5. For small expected frequencies case, the binomial test should be used. For more than two categories case, the expected frequency in any category should not be less than 1 and is not allowed to have more than 20% of the categories with expected frequencies less than 5.

5.3.1.2 Performing Chi-square Goodness-of-Fit Test Using SPSS

The chi-square test can be carried out using SPSS by following the steps below after opening the data set *"Illustration 5.1.sav"* (Figure 5.3).

Analyze → **Nonparametric Tests** → **Legacy Dialogs** → **Chi − square**

- The **Chi-square Test** dialog box will appear (as in Figure 5.4). Choose "hapmar" (Happiness of marriage) in "Test Variable List" section to test

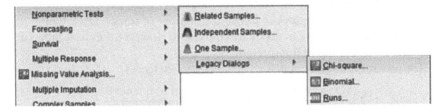

Figure 5.3 Path for chi-square goodness-of-fit test. Source: Reprint Courtesy of International Business Machines Corporation, © International Business Machines Corporation.

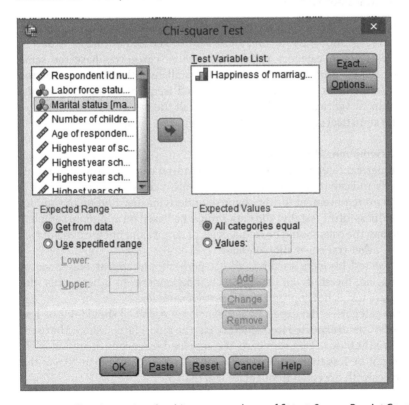

Figure 5.4 Choosing options for chi-square goodness-of-fit test. Source: Reprint Courtesy of International Business Machines Corporation, © International Business Machines Corporation.

the hypothesis that the respondents in data set are equally happy about their married life. Generally, more than one variable can be selected and a separate chi-square test will be calculated by the software. Specify the values in "Expected Values" or just choose "All categories are equal."

- *Output*: In the output table (Table 5.2), the "Expected N" represents the average score if all categories were equal; since there are three categories of happiness, the expected frequency is calculated as (1337/3 = 455.7). The residual is calculated by taking a difference of each expected frequency from the observed frequency. Based on the observed frequency and the residual, clearly there is a difference between the degree of happiness among respondents. However, the chi-square test is used to statistically test this difference.

- *Interpretation*: The testing hypothesis or the null hypothesis for the chi-square test is that the proportions in each category of the test variable are equal. The results are presented in Table 5.3. At 5% significance level, we reject the null hypothesis of equal proportions for the "Happiness of marriage" (p-value = Asymptotic significance < 0.05). This means that not all the respondents are equally happy from their marriages (rather some are very happy in fact) and the proportions across levels of happiness are statistically different.

- *Reporting the results*: A chi-square goodness-of-fit test was computed for comparing Happiness of marriage (1, Very happy; 2, Pretty happy; 3, Not too happy) in equal proportion. Statistically significant deviation from the hypothesized equal proportion was found ($\chi^2 = 750.70$, p-value < 0.05).

Table 5.2 Frequencies for the chi-square goodness-of-fit test.

	Observed N	Expected N	Residual
Very happy	855	445.7	409.3
Pretty happy	445	445.7	−0.7
Not too happy	37	445.7	−408.7
Total	1337		

Table 5.3 Chi-square goodness-of-fit test results.

	Happiness of marriage
Chi-square	750.702[a]
df	2
Asymptotic significance	0.000

a)   0 cells (0.0%) have expected frequencies less than 5.
   The minimum expected cell frequency is 445.7.

5.3.2 Testing for Independence

The chi-square test for independence (also known as a test of association) is used for testing the relationship between two categorical variables in a cross-classification contingency table. The cross-classification is used to test whether the observed patterns of the variables are dependent on each other. The chi-square test of independence is different from the goodness-of-fit test, which is based on testing the observed (experimental) and expected (theoretical) on a single variable while the test of independence seeks to test the relationship between two categorical variables. The null and alternative hypothesis for the chi-square test of independence can be written as follows:

H_0: The two variables *are independent* of each other.
H_1: The two variables are NOT independent of each other.

OR

H_0: There is no association between Variable 1 and Variable 2.
H_1: There exists an association between the two variables.

5.3.2.1 Assumptions About Data
- There must be two or more categories for each of the two categorical variables.
- The observations in each group should be independent and expected frequencies for each category should appear only once. Both the chi-square test for goodness of fit and the test of independence are not appropriate for paired samples.
- The sample size should be large enough to ensure that the expected frequency in each cell is at least 1, and the majority of cell have the expected count of at least 5.

5.3.2.2 Performing Chi-square Test of Independence Using SPSS
Consider the same data set "*Illustration 5.1.sav*," as used in the previous section. The goodness-of-fit test revealed that the distribution of responses for "Happiness of marriage" (hapmar) is not the same for all of its categories: 1 – Very happy, 2 – Pretty happy, 3 – Not too happy.

Now, let us assume that a researcher wants to test whether the "Happiness of marriage" is associated with some other variable. For instance, the interest is to test the association between "Happiness of marriage" and "General happiness." To conduct the chi-square test for association or independence, follow the stated steps:

- Open the data set "*Illustration 5.1.sav*."
- Step 2, go to: **Analyze → Descriptive Statistics → Crosstabs** (Figure 5.5).

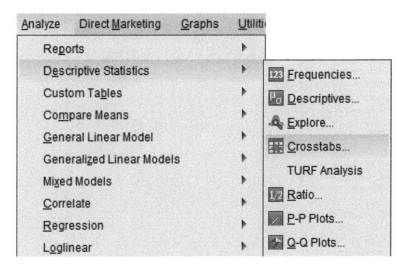

Figure 5.5 Path for the chi-square test of independence. Source: Reprint Courtesy of International Business Machines Corporation, © International Business Machines Corporation.

- The **Crosstabs** dialog box will appear (as in Figure 5.6). Since the chi-square test of independence is used to test the association between two variables, one variable should be selected in "Row(s)" and the other in "Column(s)" sections.

  In case of selecting more than one variable in either Row or Column, a separate chi-square test will be carried out for each bivariate relationship. Here, the "hapmar (Happiness of marriage)" is selected in "Row(s)" and the "happy (General happiness)" in the "Column(s)" sections. Check on the "Display clustered bar chart" option to get the visualization of the association between the variables.

- From the **Statistics** options (as in Figure 5.7), choose "Chi-square" and continue to proceed. From **Cells** options, choose the "Observed" and "Expected" counts. The "Unstandardized Residuals" can be chosen to get the output for the difference between the observed and expected frequencies.

- The first table in the output is the Crosstabs Results, where the expected and observed count and their unstandardized residuals are shown. From Table 5.4, the expected and observed counts are not the same for Happiness of marriage and General happiness. This indicates that they are not independent variables. In other words, Happiness of marriage and General happiness are related to each other.

- The second table (Table 5.5) presents the results for the chi-square test of independence. Since the p-value is less than the error margin

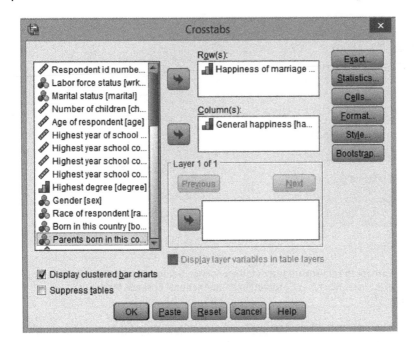

Figure 5.6 Choosing options for the chi-square test of independence – 1. Source: Reprint Courtesy of International Business Machines Corporation, © International Business Machines Corporation.

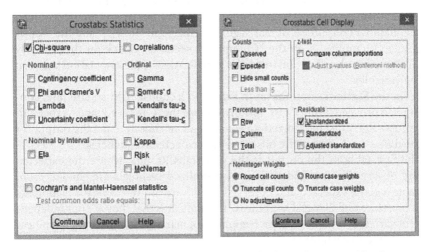

Figure 5.7 Choosing options for the chi-square test of independence – 2 and 3. Source: Reprint Courtesy of International Business Machines Corporation, © International Business Machines Corporation.

Table 5.4 Chi-square test for independence: crosstabs results.

		General happiness			Total
		Very happy	Pretty happy	Not too happy	
Panel A: Happiness of marriage					
Very happy	Count	523	310	19	852
	Expected count	370.3	436.2	45.5	852.0
	Residual	152.7	−126.2	−26.5	
Pretty happy	Count	49	357	35	441
	Expected count	191.7	225.8	23.5	441.0
	Residual	−142.7	131.2	11.5	
Not too happy	Count	6	14	17	37
	Expected count	16.1	18.9	2.0	37.0
	Residual	−10.1	−4.9	15.0	
Total	Count	578	681	71	1330
	Expected count	578.0	681.0	71.0	1330.0
Panel B: Happiness of marriage2 × General happiness crosstabulation					
Very happy	Count	523	310	19	852
	Expected count	370.3	436.2	45.5	852.0
	Residual	152.7	−126.2	−26.5	
Pretty happy	Count	55	371	52	478
	Expected count	207.7	244.8	25.5	478.0
	Residual	−152.7	126.2	26.5	
Total	Count	578	681	71	1330
	Expected count	578.0	681.0	71.0	1330.0

(p-value < 0.05), we can reject the null hypothesis of independence or no association. It can be concluded that General happiness is associated with Happiness of marriage as per the data.

The readers should notice that there is one cell with expected value less than 5 and the SPSS gives "*a. 1 cells (11.1%) have expected count less than 5. The minimum expected count is **1.98**.*" So, to avoid this problem, do the following steps:

1. For the variable "Happiness of marriage," merge the category "Not too happy" with "Pretty happy" using **Transform → Recode into Different variables**. Let's name it as "Happiness of marriage2."
2. Redo the previous steps again staring from "Step 2" after replacing "Happiness of marriage" with "Happiness of marriage2" in the analysis. One will get the following results as in Table 5.4, Panel B.

Table 5.5 Chi-square test of independence results.

	Value	df	Asymptotic significance (two-sided)
Panel A			
Pearson chi-square	424.839[a]	4	0.000
Likelihood ratio	390.298	4	0.000
Linear-by-linear association	312.491	1	0.000
Number of valid cases	1330		
Panel B			
Pearson chi-square	319.859[b]	2	0.000
Likelihood ratio	352.754	2	0.000
Linear-by-linear association	305.879	1	0.000
Number of valid cases	1330		

a) 1 cell (11.1%) has expected count less than 5. The minimum expected count is **1.98**.
b) 0 cells (0.0%) have expected count less than 5. The minimum expected count is 25.52.

Chi-square test of independence is given in Table 5.5. One can see that the p-value is less than the error margin (p-value < 0.05) and one can reject the null hypothesis of independence or no association. It can be concluded that General happiness is associated with Happiness of marriage as per the data.

- The visual display for the relationship (association) between the two variables is presented as a clustered bar chart (Figure 5.8). The General happiness is taken as a column variable and the clusters are formed with respect to the categories of row variable, i.e. Happiness of marriage. Visually, it is clear that General happiness is more for the respondents who are happy from their marriage. In other words, General happiness is significantly associated with Happiness of marriage.

5.3.3 Testing for Homogeneity

Chi-square test for homogeneity is used for testing the research hypothesis for whether two or more populations or subgroups of populations have the same distribution for a single categorical variable. Commonly, a z-test for proportions is used when the aim is to compare the variable proportions in two populations (or two subgroups); however, for testing the same in two or more populations or subgroups, the chi-square test for homogeneity is appropriate.

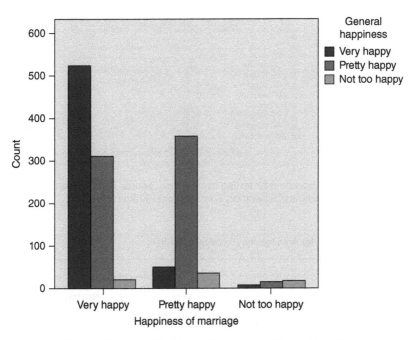

Figure 5.8 Clustered bar chart: Happiness of marriage vs. General happiness.

5.3.3.1   Assumptions About Data

- The chi-square test for homogeneity assumes that the population from which the random sample is selected is normal.
- The chi-square test can be used for the variables measured at any type of scale: nominal, ordinal, interval, or ratio.
- The expected frequency, i.e. the number of cases in each category should be at least 5 or more. The sample size should be sufficiently large to meet this assumption.

5.3.3.2   Performing Chi-square Test of Homogeneity Using SPSS

Consider the same data set "*Illustration 5.1.sav*," as used in the previous section. The first step to test the homogeneity of variance is to weight the cases based on some variable.

- Follow the path: **Data → Weight Cases**; the Weight Cases window will appear (Figure 5.9). Choose the variable "Sex" in the "Frequency Variable" to weight the cases in analysis based on gender.
- After weighting the cases, run the **Crosstabs** from **Descriptive Statistics** menu and follow the same steps as shown in Figures 5.5–5.7. The Crosstabs Results in Table 5.6 are different from that of chi-square crosstabs results for independence in Table 5.4.

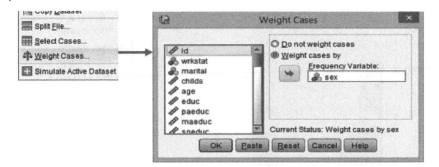

Figure 5.9 Weight cases procedure for testing homogeneity. Source: Reprint Courtesy of International Business Machines Corporation, © International Business Machines Corporation.

Table 5.6 Chi-square test for homogeneity: crosstabs results.

			Happiness of marriage			Total
			Very happy	Pretty happy	Not too happy	
General happiness	Very happy	Count	810	81	10	901
		Expected count	577.0	298.0	26.0	901.0
		Residual	233.0	−217.0	−16.0	
	Pretty happy	Count	474	540	22	1036
		Expected count	663.5	342.6	29.9	1036.0
		Residual	−189.5	197.4	−7.9	
	Not too happy	Count	27	56	27	110
		Expected count	70.4	36.4	3.2	110.0
		Residual	−43.4	19.6	23.8	
Total		Count	1311	677	59	2047
		Expected count	1311.0	677.0	59.0	2047.0

- The second table (Table 5.7) presents the results for the chi-square test of Homogeneity. Since, the p-value is less than the error margin (p-value < 0.05), the null hypothesis of homogeneity will be rejected. It can be concluded that the distribution of males and females are not homogeneous in the two variables under consideration.
- The clustered bar chart in Figure 5.8 (Section 5.3.2), where the chart shows no associations between the two variables, is different from the one in Figure 5.10 (Section 5.3.3), where one can see that the distribution of male is completely different from the distribution of female in term of happiness.

Table 5.7 Chi-square test for homogeneity: test statistic and
p-value.

	Value	df	Asymptotic significance (two-sided)
Pearson chi-square	648.232[a]	4	0.000
Likelihood ratio	590.459	4	0.000
Linear-by-linear association	479.984	1	0.000
Number of valid cases	2047		

a) 1 cell (11.1%) has expected count less than 5. The minimum
   expected count is 3.17.

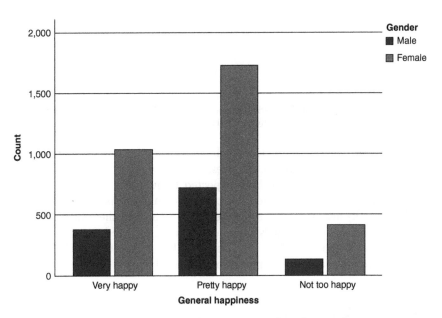

Figure 5.10 Clustered bar chart: testing for homogeneity based on gender.

5.3.4 What to Do if Assumption Violates?

If the data is not categorized numerically, it can be converted to categories from
numeric or string variables. For this, the recoding options from **Transform**
menu can be used (as in Figure 5.11). Choose "Recode into different variables,"
and then choose the variable to be saved as a new output variable and specify the categories along with their numeric values. This will be done in the

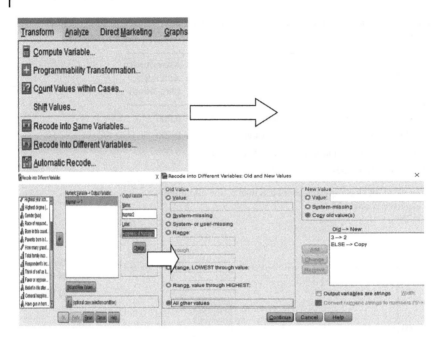

Figure 5.11 Path for transforming variables. Source: Reprint Courtesy of International Business Machines Corporation, © International Business Machines Corporation.

following same steps as we did for recoding the "Happiness of marriage" to be two categories instead of three categories.

If more than 20% of cells exhibit expected frequencies to be less than 5, the categories with low frequencies can be combined as a remedy. This problem usually arises when the sample size is small and the total responses are not enough to be adequately distributed among the predefined categories of responses. Combining the categories with lesser counts will elevate the expected frequency and will help to resolve the problem.

5.4 Mann-Whitney U Test

In nonparametric analysis, the Mann-Whitney U test is used for comparing two groups of cases on one variable. This test is an alternative to the parametric test for "two-independent samples *t*-test," which aimed to compare the means of two independent samples as discussed in Chapter 4. The Mann-Whitney U test is, however, based on mean ranks or medians, rather than the mean of original values. The null hypothesis for Mann-Whitney U test is stated as "the median score of dependent variable is same across the categories of independent variable."

5.4.1 Assumption About Data

- The data must be obtained from two independent random samples; there should be two independent categories of the independent variable to test the group differences.
- The test aims to compare the difference between two distributions of the random samples. The shape (variability) of the distribution is assumed to be the same, and only the location (central tendency) is allowed to vary across the groups.
- The dependent variable can be either ordinal or continuous but not normally distributed.

5.4.2 Mann-Whitney Test Using SPSS

The Mann-Whitney U test can be carried out using SPSS via two different ways. Let us open the data set "*Illustration 5.1.sav*" for testing if there are any statistical differences in the "Higher degree" among the male and female? This hypothesis could be tested by following the below-mentioned steps (Figure 5.12):

Method 1:

Analyze → Nonparametric Tests → Legacy Dialogs

→ 2 Independent Samples

- The "Two-Independent-Samples Tests" dialog box will appear (as in Figure 5.13). Choose the variable "degree" (Highest degree) in "Test Variable List" and specify the categories of Gender in "Groups"; Group1: 1, Group2: 2, the ordering does not matter.
- *Output*: The "Ranks" in output are important to look at the "Mean Ranks" to see which category exhibits relatively higher or lower mean ranks (see Table 5.8). The mean rank for the "males" is relatively higher than the "females," but the results of "Test Statistics" table (Table 5.9) show that this difference is statistically insignificant.

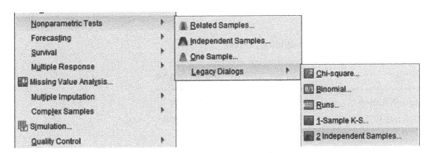

Figure 5.12 Path 2 for Mann-Whitney U test. Source: Reprint Courtesy of International Business Machines Corporation, © International Business Machines Corporation.

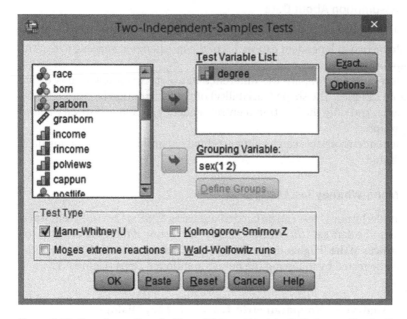

Figure 5.13 Choosing options for Mann-Whitney U test (path 1). Source: Reprint Courtesy of International Business Machines Corporation, © International Business Machines Corporation.

Table 5.8 Ranks for the variable Gender.

	Gender	N	Mean rank	Sum of ranks
Highest degree	Male	1 228	1 431.67	1 758 091.00
	Female	1 594	1 395.96	2 225 162.00
	Total	2 822		

Table 5.9 Test statistics: Mann-Whitney U test.

	Highest degree
Mann-Whitney U	953 947.000
Wilcoxon W	2 225 162.000
Z	−1.259
Asymptotic significance (two-tailed)	0.208

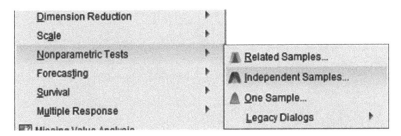

Figure 5.14 Path 2 for Mann-Whitney U test. Source: Reprint Courtesy of International Business Machines Corporation, © International Business Machines Corporation.

- *Interpretation*: At 5% significance level, we fail to reject the null hypothesis of no difference. The mean rank for the highest degree achieved is statistically the same across the categories of gender.
- *Reporting the results*: A Mann-Whitney U test indicated that the highest degree attained for males and females is the same ($U = 953\,947$, p-value $= 0.208$).

Method 2: Follow the below-given steps for the second method of performing the Mann-Whitney U test in SPSS for the same data set *"Illustration 5.1.sav"* (Figure 5.14):

Analyze → Nonparametric Tests → Independent Samples

- In "Objectives" tab, choose "Customize analysis" (as in Figure 5.15).
- In "Fields" tab, choose the variable "degree" (Highest degree) as a "Test Field" and "Gender" as "Groups" (as in Figure 5.16).
- In "Settings" tab, choose the option "Customize tests" and check on "Mann-Whitney U (2 samples)," and click "Run" (as in Figure 5.17).
- *Output and interpretation*: This method gives a detailed output as compared with the one given by "Method 1." The "Hypothesis test summary" (in Table 5.10) suggests retaining the null hypothesis of no statistical difference.
- *Detailed output*: Double-click on the "Hypothesis Test Summary" (shown in Table 5.10) in the output window to open the "Model Viewer" window. The Model Viewer allows to look at the output from different views. The distribution graph (Figure 5.18) shows that the shape of approximate distribution of "Highest degree rank score" is almost the same for male and female.

5.4.3 What to Do if Assumption Violates?

- If the samples are not independent, the Wilcoxon Signed-Rank test should be used for related or matched samples.
- The continuous variables can be transformed to categorical variables to compare the group differences for a test variable to run the test.

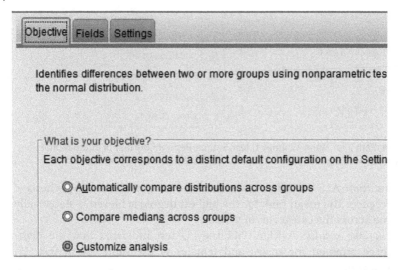

Figure 5.15 Mann-Whitney U test: option 1, path 2. Source: Reprint Courtesy of International Business Machines Corporation, © International Business Machines Corporation.

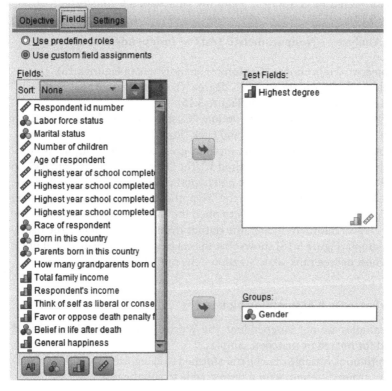

Figure 5.16 Mann-Whitney U test: option 2, path 2. Source: Reprint Courtesy of International Business Machines Corporation, © International Business Machines Corporation.

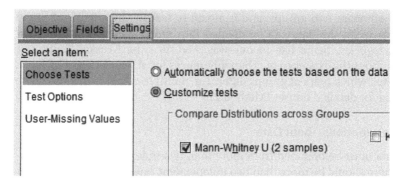

Figure 5.17 Mann-Whitney U test: option 3, path 2. Source: Reprint Courtesy of International Business Machines Corporation, © International Business Machines Corporation.

Table 5.10 Hypothesis test summary for Mann-Whitney U test.

	Null hypothesis	Test	Significance	Decision
1	This distribution of Highest degree is the same across categories of Gender	Independent-samples Mann-Whitney U test	0.208	Retain the null hypothesis

Asymptotic significances are displayed. The significance level is 0.05.

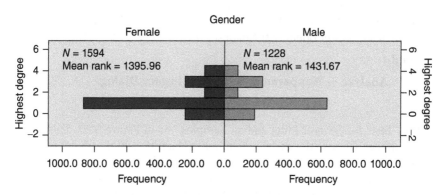

Figure 5.18 Mean ranks distribution graph for Mann-Whitney U test.

5.5 Kruskal-Wallis Test

The Kruskal-Wallis test is considered as an alternative test to the parametric one-way analysis of variance (ANOVA) for comparing more than two groups on one variable. It is an extension of the Mann-Whitney U test, where the aim is to compare two groups, and is aimed to test the hypothesis of group differences when the data can be divided into more than two independent groups/categories. Same like in one-way ANOVA, the Kruskal-Wallis test only tells about the existence of group differences for at least one group, but it does not tell which pairs of groups differ. The multiple comparisons should be performed for detailed output (Method 2).

5.5.1 Assumptions About Data

• The data must be obtained from more than two independent random samples; there should be more than two independent categories of the independent variable to test the group differences.
• The dependent variable should either be ordinal or measured at continuous level.
• The observations must be independent; the overlapping between groups is not allowed.
• The Kruskal-Wallis H test compares the difference between the distributions of each independent group. The shape (variability) of the distribution is assumed to be the same, and only the location (central tendency) can vary across the groups.

5.5.2 Kruskal-Wallis H Test Using SPSS

The Kruskal-Wallis H test can be carried out using SPSS for testing if there is any statistical difference in the "Happiness of marriage" scores by "Race of respondent." This hypothesis could be tested by following the below-mentioned steps after opening the data set *"Illustration 5.1.sav"* (Figure 5.19).

Method 1:

> **Analyze → Nonparametric Tests → Legacy Dialogs**
> **→ K Independent Samples**

• The "Tests for Several Independent Samples," as in Figure 5.20, dialog box will appear. Choose the variable "hapmar" (Happiness of marriage) in "Test Variable List," which is measured at "ordinal" level. Specify the categories of "race" (Race of respondent) in "Grouping variable" as Minimum: 1, Maximum: 3 since there are three categories (White, Black, and Other).
• *Output*: the "Mean Ranks" for "Happiness of marriage" (in Table 5.11) are relatively higher for "Black" as compared with the other two "race" categories. By comparing the overall mean ranks for all three categories, a slight difference between "Black" and "Other" is observed, but the mean rank for "White" is far away from that of "Black." These ranks however cannot be interpreted in statistical terms without looking at the significance test statistic and its *p*-value.

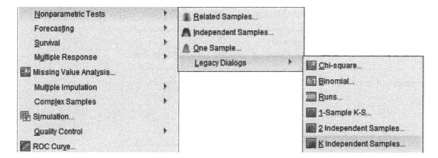

Figure 5.19 Path for Kruskal-Wallis H test (path 1). Source: Reprint Courtesy of International Business Machines Corporation, © International Business Machines Corporation.

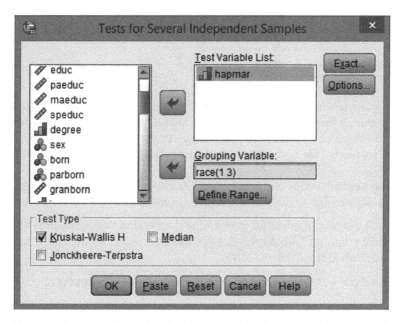

Figure 5.20 Options for Kruskal-Wallis H test (path 1). Source: Reprint Courtesy of International Business Machines Corporation, © International Business Machines Corporation.

- *Interpretation*: At 5% significance level, we reject the null hypothesis of no statistical significance difference. The mean rank for the "Happiness of marriage" is statistically different for at least one category of race. The results however do not show which categories are statistically different (refer to Table 5.12).

Table 5.11 Ranks for the variable Happiness of marriage.

	Race of respondent	N	Mean rank
Happiness of marriage	White	1143	653.56
	Black	121	785.74
	Other	73	717.23
	Total	1337	

Table 5.12 Kruskal-Wallis test results.

	Happiness of marriage
Chi-square	19.996
df	2
Asymptotic significance	0.000

- *Reporting the results*: A Kruskal-Wallis test was conducted to evaluate differences among three races (White, Black, Others) on mean ranks change in the level of Happiness of marriage (Very happy, Pretty happy, Not too happy). The test (corrected for tied ranks) was statistically significant ($\chi^2 = 19.996$, $p = 0.000$).
  Note: The Kruskal-Wallis test is calculated by chi-square statistic; however, the value for Kruskal-Wallis statistic is given in Method 2.

Method 2: The second method for performing the Kruskal-Wallis test in SPSS for the same data set *"Illustration 5.1.sav"* can be applied by following the below-mentioned steps (Figure 5.21):

Analyze → Nonparametric Tests → Independent Samples

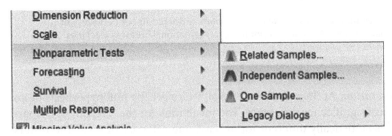

Figure 5.21 Path for Kruskal-Wallis H test (path 2). Source: Reprint Courtesy of International Business Machines Corporation, © International Business Machines Corporation.

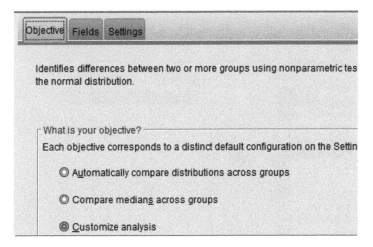

Figure 5.22 Options for Kruskal-Wallis H test (path 2). Source: Reprint Courtesy of International Business Machines Corporation, © International Business Machines Corporation.

- In "Objectives" tab, choose "Customize analysis" (Figure 5.22).
- In "Fields" tab, choose the variable "hapmar" (Happiness of marriage) as a "Test Field" and "race" (Race of respondents) as "Groups" (Figure 5.23).
- In "Settings" tab, choose the option "Customize tests" and check on "Kruskal-Wallis 1-way ANOVA (k samples)" with "All pair wise" multiple comparisons option. Click "Run" to get the output (Figure 5.24).
- *Output and interpretation*: This method gives the detailed output as compared with the one given by "Method 1." The "Hypothesis test summary" (in Table 5.13) suggests rejecting the null hypothesis of no difference.
- *Detailed output*: Double-click on the "Hypothesis Test Summary" table (as in Table 5.13) in the output window to open the "Model Viewer" window. The Model Viewer allows to look at the output from different views (Table 5.14). Choose the "pair wise comparisons" to map the group differences (Table 5.15). The pair-wise comparisons exhibit significant differences between the "White" and "Black" race in terms of their "Happiness of marriage" (see Figure 5.25). The highest score here tends toward lesser happiness, and the graph shows that Whites are happier than Blacks from their marriages.
- *Reporting the results*: A Kruskal-Wallis test was conducted to evaluate differences among three races (White, Black, Others) on mean ranks change in the scores of Happiness of marriage. One can see that, the p-value (adjusted significance) = 0.000 for White–Black, which indicates that there is statistical significance between these two categories.

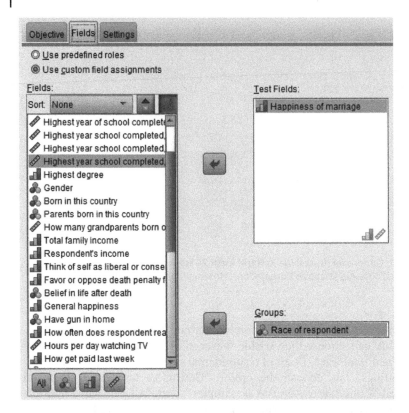

Figure 5.23 Kruskal-Wallis test: option 1, path 2. Source: Reprint Courtesy of International Business Machines Corporation, © International Business Machines Corporation.

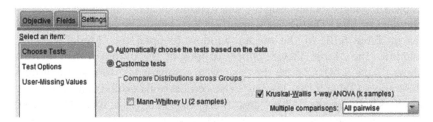

Figure 5.24 Kruskal-Wallis test: option 2, path 2. Source: Reprint Courtesy of International Business Machines Corporation, © International Business Machines Corporation.

5.5.3 Dealing with Data When Assumption Is Violated

- If the samples are not independent (rather they are related), then the Friedman's, Kendell's, or Cochran's test for related samples should be used instead of the Kruskal-Wallis test.

Table 5.13 Hypothesis test summary for Kruskal-Wallis H test.

	Null hypothesis	Test	Significance	Decision
1	This distribution of Happiness of marriage is the same across categories of Race of respondent	Independent-samples Kruskal-Wallis test	0.000	Reject the null hypothesis

Asymptotic significances are displayed. The significance level is 0.05.

Table 5.14 Independent samples test view.

Total N	2822
Test statistic	1.585
Degrees of freedom	1
Asymptotic significance (two-sided test)	0.208

Table 5.15 Sample average rank of Race of respondents.

Sample 1– Sample 2	Test statistic	Standard error	Standard test statistic	Significance	Adj. Sig. (p-value)
White–Other	−63.672	39.042	−1.631	0.103	0.309
White–Black	−132.183	30.917	−4.275	0.000	0.000
Other–Black	68.511	47.928	1.429	0.153	0.459

Each row tests the null hypothesis that Sample 1 and Sample 2 distributions are the same. Asymptotic significances (two-sided tests) are displayed. The significance level is 0.05.

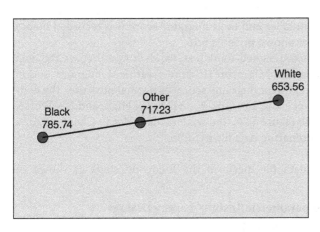

Figure 5.25 Pairwise comparisons of Race of respondents.

5.6 Wilcoxon Signed-Rank Test

The Wilcoxon Signed-Rank test is an alternative test to the parametric "Paired-samples T-Test" to test the statistical differences in the mean between two related/dependent random samples. The paired samples can come from either the same participants or the participants with similar characteristics. For instance, the effect of treatment for cancer can be tested by taking blood samples of cancer patients before and after going through the procedure. Or the effect of certain medicine can be tested for the patients randomly assigned to treatment and control groups.

5.6.1 Assumptions About Data

- The independent variable should be categorical, consisting of two "related/dependent" or "matched" groups.
- The dependent variable should be either ordinal or measured at continuous/scale level.
- The distribution is assumed to be symmetrical for the two related groups, where the two distributions have similar shape (variability) and only the location (central tendency) can vary across the groups.

5.6.2 Wilcoxon Signed-Rank Test Using SPSS

Illustration 5.2 Assume a researcher wants to see whether coffee and mild pain killer is effective for migraine headaches. Consider a hypothetical data for 30 respondents ranking the pain for their migraine headache on a 5-point Likert scale ranging from "Very low" to "Worse" (the data is generated by random assignment of migraine scale from 1 to 5). Each of them was given a cup of coffee with a mild pain killer and their migraine score was recorded before and after the treatment as shown in Table 5.16.

To carry out the Wilcoxon Signed-Rank test, we shall first prepare the data file in SPSS by defining the Mig_score1(Before treatment migraine score) and Mig_score2(After treatment migraine score) as ordinal measure. The coding was defined as 1 – Very low, 2 – Low, 3 – Mild, 4 – High, and 5 – Worse. The data file will look like Figure 5.26. The readers may refer to Chapter 2 for a detailed procedure on preparing data file in SPSS.

After preparing the data file, perform the following steps as shown in Figure 5.27:

> **Analyze → Nonparametric Tests → Legacy Dialogs**
> **→ 2 Related Samples**

Table 5.16 Migraine score before and after treatment.

Migraine score	
Before treatment	**After treatment**
3.0	2.0
3.0	1.0
5.0	3.0
1.0	1.0
4.0	4.0
2.0	3.0
4.0	4.0
5.0	2.0
5.0	3.0
4.0	1.0
5.0	3.0
1.0	4.0
5.0	3.0
3.0	1.0
2.0	2.0
4.0	4.0
3.0	3.0
3.0	1.0
5.0	3.0
1.0	5.0
3.0	1.0
1.0	2.0
3.0	1.0
4.0	3.0
4.0	4.0
4.0	5.0
5.0	1.0
3.0	3.0
1.0	3.0
5.0	2.0

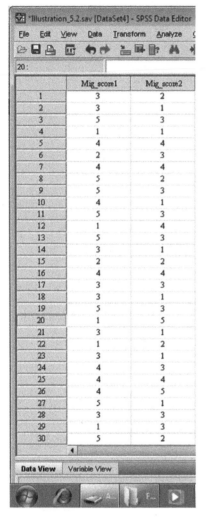

Figure 5.26 Data file in SPSS for Wilcoxon Signed-Rank test. Source: Reprint Courtesy of International Business Machines Corporation, © International Business Machines Corporation.

- The "Two-Related-Samples Tests" dialog box will appear (as in Figure 5.28). Choose the "Before Treatment Migraine Score" and "After Treatment Migraine Score" in "Variable 1" and "Variable 2." Check on "Wilcoxon" and click "OK."
- *Interpretation*: The "Ranks" output (Table 5.17) shows that the mean rank of scores after treatment of migraine is relatively lower than that of before treatment. For eight of the cases, the migraine scores were indifferent for the treatment given.
- At 5% significance level, we reject the null hypothesis of no difference (Table 5.18). The score of migraine headache turned out to be statistically different before and after the treatment was given.

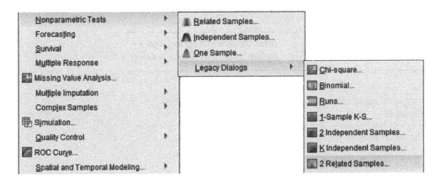

Figure 5.27 Path for Wilcoxon Signed-Rank test. Source: Reprint Courtesy of International Business Machines Corporation, © International Business Machines Corporation.

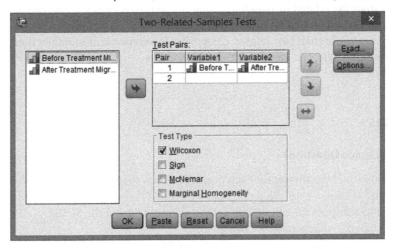

Figure 5.28 Options for Wilcoxon Signed-Rank test. Source: Reprint Courtesy of International Business Machines Corporation, © International Business Machines Corporation.

Table 5.17 Ranks for the before and after treatment scores.

		N	Mean rank	Sum of ranks
After treatment migraine score – Before treatment migraine score	Negative ranks	16[a]	12.06	193.00
	Positive ranks	6[b]	10.00	60.00
	Ties	8[c]		
	Total	30		

a) After treatment migraine score < Before treatment migraine score.
b) After treatment migraine score > Before treatment migraine score.
c) After treatment migraine score = Before treatment migraine score.

Table 5.18 Results for the Wilcoxon Signed-Rank test.

	After treatment migraine score − Before treatment migraine score
Z	−2.196[a]
Asymptotic significance (two-tailed)	0.028

a) Based on positive ranks.

5.6.3 Remedy if Assumption Violates

- If the two categories of independent variable consist of independent samples (rather than the related or matched), then the Mann-Whitney U test can be used to test the group differences.
- If there are more than two related categories of independent variable, then the Friedman test can be used as an alternative to parametric two-way ANOVA.

Exercises

Multiple-Choice Questions

Note: Choose the best answer for each question. More than one answer may be correct.

1. How to compute the chi-square cross-table value using the SPSS?
   a. Analyze → Crosstabs → Chi-square
   b. Analyze → Nonparametric → Tests → Chi-square
   c. Analyze → Descriptive statistics → Crosstabs → Statistics
   d. Analyze → Descriptive statistics → Crosstabs → Chi-square

2. The nonparametric tests can be regarded as
   a. A part of parametric tests
   b. Assumption free
   c. Distribution free
   d. Both b and c

3. Common assumptions in nonparametric tests are
   a. Randomness
   b. Independence
   c. Normality
   d. Homogeneity

4. The chi-square test can NOT be used as a
   a. Goodness-of-fit test
   b. Test of association
   c. Measure for comparison of means
   d. Test for homogeneity

5. The Mann-Whitney U test is used to
   a. Compare one group mean to a specified value
   b. Compare group variances
   c. Compare means for two related groups
   d. Compare means for two independent groups

6. The Wilcoxon Signed-Rank test is used to
   a. Compare one mean to a specified value
   b. Compare group variances
   c. Compare means for two related groups
   d. Compare means for two independent groups

7. The Kruskal-Wallis test is used to
   a. Compare one mean to a specified value
   b. Compare group variances
   c. Compare means for two independent groups
   d. Compare means for more than two groups

8. If the samples are not independent (rather they are related), then following tests can be used instead of Kruskal-Wallis test.
   a. One-way ANOVA
   b. Friedman's test
   c. Kendell's test
   d. Mann-Whitney U test

9. To test the randomness, following test is available in SPSS:
   a. Runs test
   b. Chi-square test
   c. ANOVA
   d. None of the above

10. The paired sample refers to
    a. The responses obtained from similar group of people
    b. The responses obtained from different group of people
    c. The responses obtained from same group of people
    d. All of the above

Short-Answer Questions

1. Discuss the common assumptions in using the nonparametric tests and means for testing them.

2. What do you mean by goodness of fit? Discuss its application in research.

3. What is the difference between Mann-Whitney test and Kruskal-Wallis test? In a situation where Kruskal-Wallis test is appropriate, can Mann-Whitney be used? If so, what will be the repercussion?

4. Discuss the application of chi-square in research.

5. What benefits in terms of internal and external validity and otherwise will you get in random selection of subjects?

Answers

Multiple-Choice Questions

1. c

2. c

3. a, b

4. c

5. d

6. c

7. d

8. b

9. a

10. c

6

Assumptions in Nonparametric Correlations

6.1   Introduction

Nonparametric tests are generally used where the data does not follow a normal distribution. Besides the violation of normality, the data type is also an important consideration while choosing an appropriate statistical method. For instance, the data in hand may be based on ordinal or nominal ratings (categorical data) rather than originally measured data observations at a continuous scale (interval or ratio). To deal with such data, the nonparametric counterparts are used. Specifically, for testing the bivariate relationships among variables, several measures are available for correlations and association. The conventional bivariate Pearson's moment correlation requires the data to be measured at interval or ratio scale; however, for categorical data, its extensions are available, which are Spearman's rank-order correlation, Phi coefficient, and the point-biserial correlations. These can be considered as special cases to the traditional Pearson's correlation coefficient and sometimes can be regarded as nonparametric measures for correlation and association.

6.2   Spearman Rank-Order Correlation

It is a nonparametric statistical technique for measuring the relationship between two ordinal variables or rank-ordinal correlation. If the sample size is greater than or equal to 4, then the Spearman rank-order correlation is computed as follows:

$$R_s = 1 - \frac{6 \sum_{i=1}^{n} D_i^2}{n(n^2 - 1)}$$

where

n is the number of rank pairs and D_i is the difference between a ranked pair; $D_i = R_{Xi} - R_{Yi}$ represents the difference in ranks for each observation.

Testing Statistical Assumptions in Research, First Edition. J. P. Verma and Abdel-Salam G. Abdel-Salam.
© 2019 John Wiley & Sons, Inc. Published 2019 by John Wiley & Sons, Inc.
Companion Website: www.wiley.com/go/Verma/Testing_Statistical_Assumptions_Research

Illustration 6.1   Nine women were involved in a study to examine the resting heart rate regarding frequency of visits to the gym. The assumption is that the person who visits the gym more frequently for a workout will have a slower heart rate. Table 6.1 shows the number of visits each participant made to the gym during the month of the study. It also provides the mean heart rate measured at the end of the week during the final three weeks of the month.

To compute the Spearman rank-order correlation, data file in IBM SPSS[®1] Statistics software (SPSS) needs to be prepared by defining the variables *Subject, Number,* and *Heart_Rate* as scale variables. The data file will look like Figure 6.1.

Follow the path: **Analyze → Correlate → Bivariate** as shown in Figure 6.2 to get the Bivariate Correlations window (Figure 6.3).

To estimate the Spearman correlation, choose the variables *Number* and *Heart_Rate* and check on "Spearman" correlation. Click **OK** to get the output.

- *Interpretation*: The Spearman correlation (r_s) is -0.513 (see Table 6.2). The absolute value is to be used instead $(r_s = 0.513)$; there is a moderate correlation between the number of visits and mean heart rate. However, the *p*-value suggests that this correlation is statistically insignificant at 5% significance level.

Table 6.1 Heart rate and number of visits.

Subjects	Number of visits (X)	Mean heart rate (Y)
1	7	105
2	11	110
3	10	90
4	8	89
5	9	88
6	12	78
7	9	96
8	11	76
9	18	80

---

1 SPSS Inc. was acquired by IBM in October 2009.

	Subject	Number	Heart_Rate	var
1	1.00	7.00	105.00	
2	2.00	11.00	110.00	
3	3.00	10.00	90.00	
4	4.00	8.00	89.00	
5	5.00	9.00	88.00	
6	6.00	12.00	78.00	
7	7.00	9.00	96.00	
8	8.00	11.00	76.00	
9	9.00	18.00	80.00	

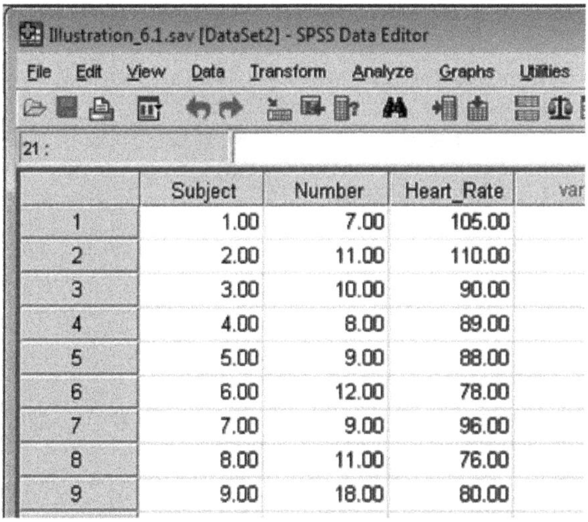

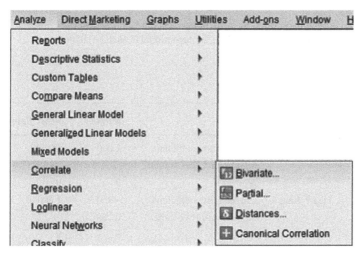

Figure 6.1 Data file in SPSS for computing Spearman rank-order correlation. Source: Reprint Courtesy of International Business Machines Corporation, © International Business Machines Corporation.

Figure 6.2 Path for bivariate correlations. Source: Reprint Courtesy of International Business Machines Corporation, © International Business Machines Corporation.

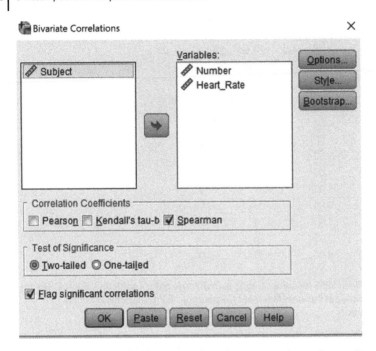

Figure 6.3 Window for defining option for Spearman correlation. Source: Reprint Courtesy of International Business Machines Corporation, © International Business Machines Corporation.

Table 6.2 Correlations.

			Number	Heart_Rate
Spearman's rho	Number	Correlation coefficient	1.000	−0.513
		Significance (two-tailed)		0.158
		N	9	9
	Heart_Rate	Correlation coefficient	−0.513	1.000
		Significance (two-tailed)	0.158	
		N	9	9

6.3   Biserial Correlation

The biserial correlations can be of two types:

- *Point-biserial correlation*: This is used when one of the variables is expressed as interval/ratio data and the other one is dichotomous nominal/categorical, i.e. with two categories. The examples of such dichotomous variables can be male/female, employed/unemployed, and so on. The point-biserial correlation uses the same computational formula, as used for the Pearson's correlation (r_{xy}).

$$r_{pb} = r_{xy} = \frac{\Sigma XY - \frac{(\Sigma X)(\Sigma Y)}{n}}{\sqrt{\left[\Sigma X^2 - \frac{(\Sigma X)^2}{n}\right]\left[\Sigma Y^2 - \frac{(\Sigma Y)^2}{n}\right]}}$$

An alternative formula using the averages ($\overline{Y}_0$, $\overline{Y}_1$) and proportions (p_0, p_1) on dichotomous variable with categories (0, 1), i.e. 0 for male and 1 for female, is given by

$$r_{pb} = \left[\frac{\overline{Y}_1 - \overline{Y}_0}{\tilde{s}_Y}\right]\sqrt{p_0 p_1}\sqrt{\frac{n}{n-1}}$$

where n is the sample size and $\tilde{S}_Y = \sqrt{\frac{\Sigma(y_i - \overline{Y})^2}{n-1}}$.

- *The biserial correlation*: The biserial correlation is used if both variables are measured on an interval/ratio scale, but one of the variables are transformed into a dichotomous variable having two categories (male/female or smoker/nonsmoker or success/failure). It provides an estimate of the value for the Pearson's correlation (based on ratio/scale data). An example of such data can be given as of the exam scores at continuous scale, but the status Pass/Fail is dichotomous (based on the scores).

Remarks

1) If the two variables are dichotomous and rank-ordinal, then Spearman rank-order correlation coefficient should be utilized instead of biserial correlation.

2) If the dichotomous data is indeed binary (i.e. success/failure) or originally based on interval/ratio scale, the point biserial correlation is appropriate; otherwise, the biserial correlation can be used.

Illustration 6.2   A data of 10 persons on the handedness and STAT test are shown in Table 6.3. The variable X (dichotomous) represents the characteristics whether a person is left-handed or right-handed and the Y variable (interval/ratio) represents the scores on STAT test. The data needs to be entered and coded in SPSS to be used for the analysis.

To prepare the data file in SPSS, we shall define the variables X (Handedness) as Nominal and Y (Test_score) as Scale, respectively. After defining the code for Handedness as 0 – Left_Handed and 1 – Right_Handed, the data file in SPSS will look like Figure 6.4.

Table 6.3 Test score and handedness data.

Subjects	Handedness (X)	Test score (Y)
1	1	11
2	1	1
3	1	0
4	1	2
5	1	0
6	0	11
7	0	11
8	0	5
9	0	8
10	0	4

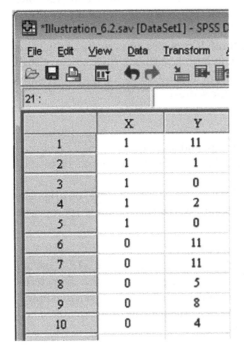

Figure 6.4 Data file in SPSS for biserial correlation. Source: Reprint Courtesy of International Business Machines Corporation, © International Business Machines Corporation.

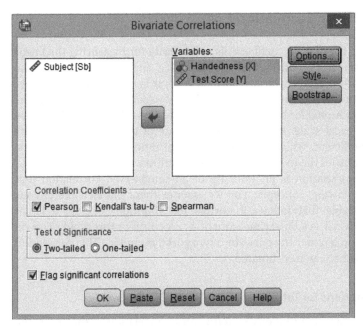

Figure 6.5 Options for bivariate correlations. Source: Reprint Courtesy of International Business Machines Corporation, © International Business Machines Corporation.

In the data view, follow the command sequence: **Analyze → Correlate → Bivariate** as shown in Figure 6.2 to get the bivariate correlation window as shown in Figure 6.5.

To estimate the point-biserial correlation, choose the variables Handedness (X) and the Test Score (Y) and check on "Pearson" correlation. Click on **OK** to get the output.

- *Interpretation*: The point-biserial correlation (r_{pb}) is -0.57 (see Table 6.4). Since the variable "Handedness" is dichotomized as 0 or 1, without ordering (no category is defined to be superior than the other), the sign of the correlation (r_{pb}) can be ignored. The absolute value is to be used instead ($r_{pb} = 0.57$); there is a moderate correlation between the handwriting and test scores on statistics. However, the p-value suggests that this correlation is statistically insignificant at 5% significance level.

Table 6.4 Correlations.

		Test score
Handedness	Pearson correlation	−0.570
	Significance (two-tailed)	0.085
	N	10

6.4 Tetrachoric Correlation

The tetrachoric correlation coefficient is appropriate for measuring the linear association between two dichotomous variables (X and Y) assuming that the underlying distribution is bivariate normal. Also, it is computed for the data measured on an interval/ratio scale that follow normal distributions, but the readings (observations) for both variables have been transformed to the binary nominal/categorical scale. It provides an estimate of the value for Pearson's correlation coefficient, which is used for interval/ratio scale data, where the original data satisfies the measurement scale condition.

Tetrachoric correlation is an estimate of an unobserved correlation that depends on following concepts: For X, assume there exists a continuous underlying variable that follows a normal distribution. Consider a latent variable; call it X_L. X is a binary measure of X_L. A cutoff or threshold point on X_L divides the area under the curve into two parts, p_x and q_x. This correlation is unobservable but can be estimated.

6.4.1 Assumptions for Tetrachoric Correlation Coefficient

- One of the main assumptions of tetrachoric correlation is that both the variables are continuous and normally distributed.
- Another assumption in using the tetrachoric correlation is that both the variables should be linearly related.

Illustration 6.3 In a study, 100 patients were classified for neurosis, based on the responses from two raters, A and B. The aim of the study was to determine whether the responses of the two raters A and B correlate with each other. Rater's responses have been shown in Table 6.5.

Tetrachoric correlation can be calculated using formula (6.1), where a, b, c, and d are the cell frequencies. The cosine formula for computing tetrachoric correlation provides a very close approximation to r_t only when both the

Table 6.5 Responses of the raters to classify subjects having neurosis.

		Rater 1		
		Neurosis diagnosed	Neurosis not diagnosed	Total
Rater 2	Neurosis diagnosed	35(a)	15(b)	50
	Neurosis not diagnosed	10(c)	40(d)	50
	Total	45	55	$N = 100$

variables are dichotomized at their medians.

$$r_t = \cos \frac{180\sqrt{bc}}{\sqrt{ad} + \sqrt{bc}} \tag{6.1}$$

Substituting the cell frequencies in the formula, we get

$$r_t = \cos \frac{180\sqrt{15 \times 10}}{\sqrt{35 \times 40} + \sqrt{15 \times 10}}$$

$$= \cos \ 44.38$$

$$= 0.72 \text{ (from Table A.1)}$$

6.4.1.1 Testing Significance

Let us see whether the value of tetrachoric correlation obtained in this illustration is significant at 5% level. We shall use the following t-test to do that:

$$t = \frac{r_t}{\sigma_{r_t}} \tag{6.2}$$

where σ_{r_t} is the standard error of r_t and is computed using the formula

$$\sigma_{r_t} = \frac{\sqrt{pqp'q'}}{uu'\sqrt{N}}$$

where p and q are the proportions of the neurosis diagnosed and neurosis not diagnosed by Rater 1, whereas p' and q' are the proportions of the neurosis diagnosed and neurosis not diagnosed by Rater 2.

Thus,

$$p = \frac{45}{100} = 0.45 \quad \text{and} \quad q = \frac{55}{100} = 0.55, \quad u = 0.3958$$

$$p' = \frac{50}{100} = 0.50 \quad \text{and} \quad q' = \frac{50}{100} = 0.50, \quad u' = 0.3989$$

The values of u and u' have been obtained from the Table A.5 corresponding to the larger area.

Remark u is the ordinate of the standard normal curve at the point cutting off a tail of that distribution with an area equal to q.

Substituting these values, we get

$$\sigma_{r_t} = \frac{\sqrt{pqp'q'}}{uu'\sqrt{N}}$$

$$= \frac{\sqrt{0.45 \times 0.55 \times 0.5 \times 0.5}}{0.3958 \times 0.3989 \times \sqrt{100}}$$

$$= 0.15$$

Thus,

$$t = \frac{r_t}{\sigma_{r_t}}$$

$$= \frac{0.72}{0.15} = 4.8$$

From Table A.2, the value of $t_{0.05,(100-2)} = t_{0.05,98} = 1.987$.
Since the value of t ($= 4.8$) is greater than 1.987, the null hypothesis may be rejected at 5% level.

Inference: Since the null hypothesis has been rejected, it may be concluded that the obtained value of tetrachoric correlation is significant. In other words, it can be concluded that there is statistically significant strong positive correlation between the raters' agreement over neurosis diagnosis of patients.

6.5   Phi Coefficient (Φ)

The Phi coefficient is suitable for calculating the correlation between two dichotomous variables in a 2×2 cross-tabulation setting. It is also used as a measure of association for the chi-square test for 2×2 tables and can be employed in physiological testing to evaluate the consistency of responses from some respondents on two questions. It takes the value of 0 when there is no association, which would be indicated by a chi-square value of 0 as well. When the variables are perfectly associated, phi assumes the value of 1 and all the observations fall just on the main or minor diagonal.

Illustration 6.4   In a study, 30 students were classified for depression on the basis of clinical examinations by two psychotherapists A and B. The data so obtained are shown in Table 6.6. On the basis of the data, can it be concluded that the responses of the two psychotherapists are correlated?

We shall first prepare the data file in SPSS by defining the two variables *Psychotherapist_A* (diagnosis result of psychotherapist A) and *Psychotherapist_B* (diagnosis result of psychotherapist B) as Nominal. Coding for both the variables are defined as 1 – depression diagnosed and 0 – depression not diagnosed.

After preparing the data file, in **Data View** menu, click on the following sequence of commands as shown in Figure 6.6 for calculating Phi coefficient in SPSS.

Analyze → Descriptive Statistics → Crosstabs

From the **Crosstabs** menu (as in Figure 6.7), choose the variables in "Row(s)" and "Column(s)" to proceed for the estimation for Phi coefficient. Click on **Statistics** tab to get Figure 6.8 to check "Phi and Cramer's V" option. After

Table 6.6 Clinical findings of psychotherapists on depression.

S. No.	Psychotherapist A	Psychotherapist B
1	1	1
2	1	1
3	1	1
4	1	0
5	0	1
6	0	0
7	1	0
8	0	0
9	0	0
10	0	1
11	1	1
12	0	0
13	0	0
14	0	0
15	1	1
16	0	0
17	1	0
18	0	0
19	1	1
20	1	1
21	0	1
22	0	0
23	0	0
24	1	0
25	0	0
26	1	1
27	1	0
28	1	1
29	1	0
30	1	1

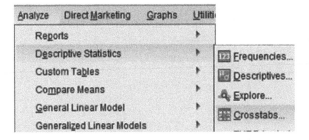

Figure 6.6 Path for crosstabs: Phi coefficient. Source: Reprint Courtesy of International Business Machines Corporation, © International Business Machines Corporation.

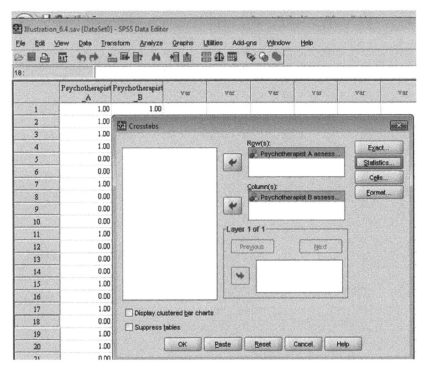

Figure 6.7 Screen showing option for selecting rows and columns in crosstabs. Source: Reprint Courtesy of International Business Machines Corporation, © International Business Machines Corporation.

pressing on Continue and OK, the results as shown in Table 6.7 shall be generated.

Interpreting the output: The Phi Cramer's *V* coefficient is significant at 5% level of significance, with a correlation coefficient of 0.413 (moderate) (see Table 6.7). It can be concluded that based on the Phi and Cramer's coefficient of correlation, there exists a statistically significant moderate positive correlation between psychotherapists' agreements.

Figure 6.8 Options for the Phi coefficient. Source: Reprint Courtesy of International Business Machines Corporation, © International Business Machines Corporation.

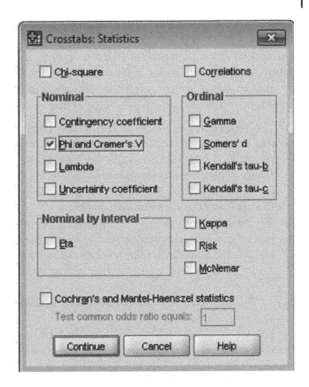

Table 6.7 Results: crosstabulation (Rater A × Rater B) and Phi coefficient.

Count: **Psychotherapist A × Psychotherapist B** *crosstabulation*

		Psychotherapist B assessment		
		Depression not diagnosed	Depression diagnosed	Total
Psychotherapist A assessment	Depression not diagnosed	11	3	14
	Depression diagnosed	6	10	16
Total		17	13	30

Symmetric measures

		Value	Approximate significance
Nominal by nominal	Phi	0.413	0.024
	Cramer's *V*	0.413	0.024
Number of valid cases		30	

6.6 Assumptions About Data

The special cases discussed for Pearson's correlation relax the assumption for using continuous variables only; rather the dichotomous variables can also be used to estimate the correlation. However, there are certain data-related assumptions associated with the methods discussed.

(a) The tetrachoric and biserial correlation coefficients assume a hypothetical underlying distribution of the data to be normal, where the data is recorded at interval/ratio scale. Because of the high dependence on normality assumption, these methods should only be applied if an empirical evidence of normal distribution is present. The Phi coefficient however does not assume a normal distribution.

(b) The relationship between two variables is assumed to be linear. However, if this relationship is nonlinear, the value of correlation coefficient could be just used as an indication only for the strength of relationship.

6.7 What if the Assumptions Are Violated?

In case, we have one variable that is quantitative and the other that is binary, then the product moment correlation is called *point-biserial correlation*, which is analog of tetrachoric. It is simply the *biserial correlation*.

If the variables are ordinal, then the most appropriate product moment correlation will be the "Spearman rank correlation coefficient." Inferred process correlation is polychoric correlation.

Exercises

Multiple-Choice Questions

Note: Choose the best answer for each question. More than one answer may be correct.

1. When you have a 2 × 2 table for two categorical variables, for measuring the association, it is better to use
   a. Chi-square test
   b. Fisher's exact test
   c. Phi correlation
   d. Pearson correlation

2. When you have a 2 × 2 table for large sample, for testing whether a statistical significance association between the two categorical variables, we use
   a. Chi-square test

    b. Fisher's exact test
    c. Phi correlation
    d. Pearson correlation

3. To compute Phi correlation coefficient, we use
    a. Analyze → Descriptive Statistics → Frequencies
    b. Analyze → Descriptive Statistics → Descriptives
    c. Analyze → Descriptive Statistics → Explore
    d. Analyze → Descriptive Statistics → Crosstabs

4. To compute biserial correlation coefficient, we use
    a. Analyze → Descriptive Statistics → Frequencies
    b. Analyze → Descriptive Statistics → Crosstabs
    c. Analyze → Descriptive Statistics → Descriptives
    d. Analyze → Descriptive Statistics → Explore

5. To compute biserial correlation coefficient, the variables should be assumed to be of which type?
    a. Both of them should be categorical.
    b. Both of them should be quantitative.
    c. Both of them should follow a normal distribution.
    d. At least one of them should follow a normal distribution.

6. Which one of the following is a special case of the Pearson correlation coefficient?
    a. Biserial correlation
    b. Phi correlation
    c. Point-biserial correlation
    d. Tetrachoric correlations

7. Which are observed correlation coefficients?
    a. Phi correlation
    b. Biserial correlation
    c. Point-biserial correlation
    d. Tetrachoric correlations

8. Which are estimated or inferred correlation coefficients?
    a. Biserial correlation
    b. Phi correlation
    c. Point-biserial correlation
    d. Tetrachoric correlations

9. The Spearman correlation coefficient is considered to be which of the following?
   a. A parametric procedure
   b. A nonparametric procedure
   c. A semi-parametric procedure
   d. None of the above

10. To compute Spearman correlation coefficient, we use
    a. Analyze → Descriptive Statistics → Frequencies
    b. Analyze → Descriptive Statistics → Descriptives
    c. Analyze → Correlate → Partial
    d. Analyze → Correlate → Bivariate

Short-Answer Questions

1. If assumption of normality gets violated, then can rank correlation be used instead of product moment correlation? Discuss the procedure of computing rank correlation.

2. What is the distinction between biserial and point-biserial correlations? Discuss by means of an example. Can product moment correlation be used in a situation where point-biserial correlation is appropriate? If so, whether the results would differ and how much?

3. What is tetrachoric correlation? Discuss the procedure in its computing and testing significance.

4. What are the assumptions in different types of nonparametric correlations? If assumptions are violated, what you propose to do?

Answers

Multiple-Choice Questions

1. c

2. a

3. d

4. b

5. c

6. b

7. a

8. a, c, d

9. b

10. d

Appendix

Statistical Tables

Table A.1 Trigonometric function.

Angle (°)	sin	cos	tan	Angle (°)
0	0	1	0	0
1	0.0174	0.9998	0.0175	1
2	0.0349	0.9994	0.0349	2
3	0.0523	0.9986	0.0524	3
4	0.0698	0.9976	0.0699	4
5	0.0872	0.9962	0.0875	5
6	0.1045	0.9945	0.1051	6
7	0.1219	0.9926	0.1228	7
8	0.1392	0.9903	0.1405	8
9	0.1564	0.9877	0.1584	9
10	0.1736	0.9848	0.1763	10
11	0.1908	0.9816	0.1944	11
12	0.2079	0.9781	0.2126	12
13	0.2249	0.9744	0.2309	13
14	0.2419	0.9703	0.2493	14
15	0.2588	0.9659	0.2679	15
16	0.2756	0.9613	0.2867	16
17	0.2924	0.9563	0.3057	17
18	0.3090	0.9511	0.3249	18
19	0.3256	0.9455	0.3443	19
20	0.3420	0.9397	0.3640	20
21	0.3584	0.9336	0.3839	21
22	0.3746	0.9272	0.4040	22

(Continued)

Testing Statistical Assumptions in Research, First Edition. J. P. Verma and Abdel-Salam G. Abdel-Salam.
© 2019 John Wiley & Sons, Inc. Published 2019 by John Wiley & Sons, Inc.
Companion Website: www.wiley.com/go/Verma/Testing_Statistical_Assumptions_Research

Table A.1 (Continued)

Angle (°)	sin	cos	tan	Angle (°)
23	0.3907	0.9205	0.4245	23
24	0.4067	0.9135	0.4452	24
25	0.4226	0.9063	0.4663	25
26	0.4384	0.8988	0.4877	26
27	0.4540	0.8910	0.5095	27
28	0.4695	0.8829	0.5317	28
29	0.4848	0.8746	0.5543	29
30	0.5000	0.8660	0.5773	30
31	0.5150	0.8571	0.6009	31
32	0.5299	0.8480	0.6249	32
33	0.5446	0.8387	0.6494	33
34	0.5592	0.8290	0.6745	34
35	0.5736	0.8191	0.7002	35
36	0.5878	0.8090	0.7265	36
37	0.6018	0.7986	0.7535	37
38	0.6157	0.7880	0.7813	38
39	0.6293	0.7772	0.8098	39
40	0.6428	0.7660	0.8391	40
41	0.6561	0.7547	0.8693	41
42	0.6691	0.7431	0.9004	42
43	0.6820	0.7314	0.9325	43
44	0.6947	0.7193	0.9657	44
45	0.7071	0.7071	1	45
46	0.7193	0.6947	1.0355	46
47	0.7314	0.6820	1.0724	47
48	0.7431	0.6691	1.1106	48
49	0.7547	0.6561	1.1504	49
50	0.7660	0.6428	1.1918	50
51	0.7772	0.6293	1.2349	51
52	0.7880	0.6157	1.2799	52
53	0.7986	0.6018	1.3270	53
54	0.809	0.5878	1.3764	54
55	0.8191	0.5736	1.4281	55
56	0.829	0.5592	1.4826	56
57	0.8387	0.5446	1.5399	57

(Continued)

Table A.1 (Continued)

Angle (°)	sin	cos	tan	Angle (°)
58	0.848	0.5299	1.6003	58
59	0.8571	0.515	1.6643	59
60	0.866	0.5	1.7321	60
61	0.8746	0.4848	1.8040	61
62	0.8829	0.4695	1.8907	62
63	0.8910	0.454	1.9626	63
64	0.8988	0.4384	2.0503	64
65	0.9063	0.4226	2.1445	65
66	0.9135	0.4067	2.2460	66
67	0.9205	0.3907	2.3559	67
68	0.9272	0.3746	2.4751	68
69	0.9336	0.3584	2.6051	69
70	0.9397	0.342	2.7475	70
71	0.9455	0.3256	2.9042	71
72	0.9511	0.309	3.0777	72
73	0.9563	0.2924	3.2709	73
74	0.9613	0.2756	3.4874	74
75	0.9659	0.2588	3.7321	75
76	0.9703	0.2419	4.0108	76
77	0.9744	0.2249	4.3315	77
78	0.9781	0.2079	4.7046	78
79	0.9816	0.1908	5.1446	79
80	0.9848	0.1736	5.6713	80
81	0.9877	0.1564	6.3138	81
82	0.9903	0.1392	7.1154	82
83	0.9926	0.1219	8.1443	83
84	0.9945	0.1045	9.5144	84
85	0.9962	0.0872	11.430	85
86	0.9976	0.0698	14.301	86
87	0.9986	0.0523	19.081	87
88	0.9994	0.0349	28.636	88
89	0.9998	0.0174	57.290	89
90	1	0	∞	90

Table A.2 Critical values of *t*.

Degrees of freedom (df)	α for two-tailed test					
	0.2	0.1	0.05	0.02	0.01	0.001
1	3.078	6.314	12.71	31.82	63.66	636.62
2	1.886	2.920	4.303	6.965	9.925	31.599
3	1.638	2.353	3.182	4.541	5.841	12.924
4	1.533	2.132	2.776	3.747	4.604	8.610
5	1.476	2.015	2.571	3.365	4.032	6.869
6	1.440	1.943	2.447	3.143	3.707	5.959
7	1.415	1.895	2.365	2.998	3.499	5.408
8	1.397	1.860	2.306	2.896	3.355	5.041
9	1.383	1.833	2.262	2.821	3.25	4.781
10	1.372	1.812	2.228	2.764	3.169	4.587
11	1.363	1.796	2.201	2.718	3.106	4.437
12	1.356	1.782	2.179	2.681	3.055	4.318
13	1.350	1.771	2.160	2.65	3.012	4.221
14	1.345	1.761	2.145	2.624	2.977	4.140
15	1.341	1.753	2.131	2.602	2.947	4.073
16	1.337	1.746	2.120	2.583	2.921	4.015
17	1.333	1.740	2.110	2.567	2.898	3.965
18	1.330	1.734	2.101	2.552	2.878	3.922
19	1.328	1.729	2.093	2.539	2.861	3.883
20	1.325	1.725	2.086	2.528	2.845	3.850
21	1.323	1.721	2.080	2.518	2.831	3.819
22	1.321	1.717	2.074	2.508	2.819	3.792
23	1.319	1.714	2.069	2.500	2.807	3.768
24	1.318	1.711	2.064	2.492	2.797	3.745
25	1.316	1.708	2.060	2.485	2.787	3.725
26	1.315	1.706	2.056	2.479	2.779	3.707
27	1.314	1.703	2.052	2.473	2.771	3.690
28	1.313	1.701	2.048	2.467	2.763	3.674
29	1.311	1.699	2.045	2.462	2.756	3.659
30	1.310	1.697	2.042	2.457	2.750	3.646
40	1.303	1.684	2.021	2.423	2.704	3.551
60	1.296	1.671	2.000	2.390	2.660	3.460
80	1.292	1.664	1.990	2.374	2.639	3.416
100	1.290	1.660	1.984	2.364	2.626	3.390
1000	1.282	1.646	1.962	2.330	2.581	3.300
∞	1.282	1.645	1.960	2.326	2.576	3.291
	0.10	0.05	0.025	0.01	0.005	0.0005

α for one-tailed test

Table A.3 Critical values for number of runs G.

									Value of n_2										
		2	3	4	5	6	7	8	9	10	11	12	13	14	15	16	17	18	19	20
Value of n_1	2	1	1	1	1	1	1	1	1	1	1	2	2	2	2	2	2	2	2	2
		6	6	6	6	6	6	6	6	6	6	6	6	6	6	6	6	6	6	6
	3	1	1	1	1	2	2	2	2	2	2	2	2	2	3	3	3	3	3	3
		6	8	8	8	8	8	8	8	8	8	8	8	8	8	8	8	8	8	8
	4	1	1	1	2	2	2	3	3	3	3	3	3	3	3	4	4	4	4	4
		6	8	9	9	9	10	10	10	10	10	10	10	10	10	10	10	10	10	10
	5	1	1	2	2	3	3	3	3	3	4	4	4	4	4	4	4	5	5	5
		6	8	9	10	10	11	11	12	12	12	12	12	12	12	12	12	12	12	12
	6	1	2	2	3	3	3	3	4	4	4	4	5	5	5	5	5	5	6	6
		6	8	9	10	11	12	12	13	13	13	13	14	14	14	14	14	14	14	14
	7	1	2	2	3	3	3	4	4	5	5	5	5	5	6	6	6	6	6	6
		6	8	10	11	12	13	13	14	14	14	14	15	15	15	16	16	16	16	16
	8	1	2	3	3	3	4	4	5	5	5	6	6	6	6	6	7	7	7	7
		6	8	10	11	12	13	14	14	15	15	16	16	16	16	17	17	17	17	17
	9	1	2	3	3	4	4	5	5	5	6	6	6	7	7	7	7	8	8	8
		6	8	10	12	13	14	14	15	16	16	16	17	17	18	18	18	18	18	18
	10	1	2	3	3	4	5	5	5	6	6	7	7	7	7	8	8	8	8	9
		6	8	10	12	13	14	15	16	16	17	17	18	18	18	19	19	19	20	20
	11	1	2	3	4	4	5	5	6	6	7	7	7	8	8	8	9	9	9	9
		6	8	10	12	13	14	15	16	17	17	18	19	19	19	20	20	20	21	21
	12	2	2	3	4	4	5	6	6	7	7	7	8	8	8	9	9	9	10	10
		6	8	10	12	13	14	16	16	17	18	19	19	20	20	21	21	21	22	22
	13	2	2	3	4	5	5	6	6	7	7	8	8	9	9	9	10	10	10	10
		6	8	10	12	14	15	16	17	18	19	19	20	20	21	21	22	22	23	23
	14	2	2	3	4	5	5	6	7	7	8	8	9	9	9	10	10	10	11	11
		6	8	10	12	14	15	16	17	18	19	20	20	21	22	22	23	23	23	24
	15	2	3	3	4	5	6	6	7	7	8	8	9	9	10	10	11	11	11	12
		6	8	10	12	14	15	16	18	18	19	20	21	22	22	23	23	24	24	25
	16	2	3	4	4	5	6	6	7	8	8	9	9	10	10	11	11	11	12	12
		6	8	10	12	14	16	17	18	19	20	21	21	22	23	23	24	25	25	25
	17	2	3	4	4	5	6	7	7	8	9	9	10	10	11	11	11	12	12	13
		6	8	10	12	14	16	17	18	19	20	21	22	23	23	24	25	25	26	26
	18	2	3	4	5	5	6	7	8	8	9	9	10	10	11	11	12	12	13	13
		6	8	10	12	14	16	17	18	19	20	21	22	23	24	25	25	26	26	27

(Continued)

Table A.3 (Continued)

	2	3	4	5	6	7	8	9	10	11	12	13	14	15	16	17	18	19	20
									Value of n_2										
19	2	3	4	5	6	6	7	8	8	9	10	10	11	11	12	12	13	13	13
	6	8	10	12	14	16	17	18	20	21	22	23	23	24	25	26	26	27	27
20	2	3	4	5	6	6	7	8	9	9	10	10	11	12	12	13	13	13	14
	6	8	10	12	14	16	17	18	20	21	22	23	24	25	25	26	27	27	28

The entries in this table are the critical G values assuming a two-tailed test with a significance level of $\alpha = 0.05$.
The null hypothesis of randomness of groupings in a sequence of alternatives,
Source: Swed and Eisenhart (1943). Reproduced with permission of Institute of Mathematical Statistics.

Table A.4 Critical values of Wilcoxon signed-rank test (n = total number of + and − signs combined).

n	Two-tailed test		One-tailed test	
	$\alpha = 0.05$	$\alpha = 0.01$	$\alpha = 0.05$	$\alpha = 0.01$
5	—	—	0	—
6	0	—	2	—
7	2	—	3	0
8	3	0	5	1
9	5	1	8	3
10	8	3	10	5
11	10	5	13	7
12	13	7	17	9
13	17	9	21	12
14	21	12	25	15
15	25	15	30	19
16	29	19	35	23
17	34	23	41	27
18	40	27	47	32
19	46	32	53	37
20	52	37	60	43
21	58	42	67	49
22	65	48	75	55
23	73	54	83	62

Table A.4 (Continued)

n	Two-tailed test		One-tailed test	
	$\alpha = 0.05$	$\alpha = 0.01$	$\alpha = 0.05$	$\alpha = 0.01$
24	81	61	91	69
25	89	68	100	76
26	98	75	110	84
27	107	83	119	92
28	116	91	130	101
29	126	100	140	110
30	137	109	151	120

— indicates that it is not possible to get a value in the critical region.
Reject the null hypothesis if the number of the less frequent sign (*x*) is less than
or equal to the value in the table.
Source: Wilcoxon (1950). Reproduced with permission of John Wiley & Sons.

Table A.5 Standard scores (or deviates) and ordinates corresponding to
divisions of the area under the normal curve into a larger proportion (*B*)
and a smaller proportion (*C*).

B	Z	u	C
The larger area	Standard score	Ordinate	Smaller area
0.500	0.000	0.3989	0.500
0.505	0.0125	0.3989	0.495
0.510	0.0251	0.3988	0.490
0.515	0.0376	0.3987	0.485
0.520	0.0502	0.3984	0.480
0.525	0.0627	0.3982	0.475
0.530	0.0753	0.3978	0.470
0.535	0.0878	0.3974	0.465
0.540	0.1004	0.3969	0.460
0.545	0.1130	0.3964	0.455
0.550	0.1257	0.3958	0.450
0.555	0.1383	0.3951	0.445
0.560	0.1510	0.3944	0.440
0.565	0.1637	0.3936	0.435
0.570	0.1764	0.3928	0.430
0.575	0.1891	0.3919	0.425

Table A.5 (Continued)

B	Z	u	C
The larger area	Standard score	Ordinate	Smaller area
0.580	0.2019	0.3909	0.420
0.585	0.2147	0.3899	0.415
0.590	0.2275	0.3887	0.410
0.595	0.2404	0.3876	0.405
0.600	0.2533	0.3863	0.400
0.605	0.2663	0.3850	0.395
0.610	0.2793	0.3837	0.390
0.615	0.2924	0.3822	0.385
0.620	0.3055	0.3808	0.380
0.625	0.3186	0.3792	0.375
0.630	0.3319	0.3776	0.370
0.635	0.3451	0.3759	0.365
0.640	0.3585	0.3741	0.360
0.645	0.3719	0.3723	0.355
0.650	0.3853	0.3704	0.350
0.655	0.3989	0.3684	0.345
0.660	0.4125	0.3664	0.340
0.665	0.4261	0.3643	0.335
0.670	0.4399	0.3621	0.330
0.675	0.4538	0.3599	0.325
0.680	0.4677	0.3576	0.320
0.685	0.4817	0.3552	0.315
0.690	0.4959	0.3528	0.310
0.695	0.5101	0.3503	0.305
0.700	0.5244	0.3477	0.300
0.705	0.5388	0.3450	0.295
0.710	0.5534	0.3423	0.290
0.715	0.5681	0.3395	0.285
0.720	0.5828	0.3366	0.280
0.725	0.5978	0.3337	0.275
0.730	0.6128	0.3306	0.270
0.735	0.6280	0.3275	0.265
0.740	0.6433	0.3244	0.260

(Continued)

Table A.5 (Continued)

B	Z	u	C
The larger area	Standard score	Ordinate	Smaller area
0.745	0.6588	0.3211	0.255
0.750	0.6745	0.3178	0.250
0.755	0.6903	0.3144	0.245
0.760	0.7063	0.3109	0.240
0.765	0.7225	0.3073	0.235
0.770	0.7388	0.3036	0.230
0.775	0.7554	0.2999	0.225
0.780	0.7722	0.2961	0.220
0.785	0.7892	0.2922	0.215
0.790	0.8064	0.2882	0.210
0.795	0.8239	0.2841	0.205
0.800	0.8416	0.2800	0.200
0.805	0.8596	0.2757	0.195
0.810	0.8779	0.2714	0.190
0.815	0.8965	0.2669	0.185
0.820	0.9154	0.2624	0.180
0.825	0.9346	0.2578	0.175
0.830	0.9542	0.2531	0.170
0.835	0.9741	0.2482	0.165
0.840	0.9945	0.2433	0.160
0.845	1.0152	0.2383	0.155
0.850	1.0364	0.2332	0.150
0.855	1.0581	0.2279	0.145
0.860	1.0803	0.2226	0.140
0.865	1.1031	0.2171	0.135
0.870	1.1264	0.2115	0.130
0.875	1.1503	0.2059	0.125
0.880	1.1750	0.2000	0.120
0.885	1.2004	0.1941	0.115
0.890	1.2265	0.1880	0.110
0.895	1.2536	0.1818	0.105
0.900	1.2816	0.1755	0.100
0.905	1.3106	0.1690	0.095

(Continued)

Table A.5 (Continued)

B The larger area	Z Standard score	u Ordinate	C Smaller area
0.910	1.3408	0.1624	0.090
0.915	1.3722	0.1556	0.085
0.920	1.4051	0.1487	0.080
0.925	1.4395	0.1416	0.075
0.930	1.4757	0.1343	0.070
0.935	1.5141	0.1268	0.065
0.940	1.5548	0.1191	0.060
0.945	1.5982	0.1112	0.055
0.950	1.6449	0.1031	0.050
0.955	1.6954	0.0948	0.045
0.960	1.7507	0.0862	0.040
0.965	1.8119	0.0773	0.035
0.970	1.8808	0.0680	0.030
0.975	1.9600	0.0584	0.025
0.980	2.0537	0.0484	0.020
0.985	2.1701	0.0379	0.015
0.990	2.3263	0.0267	0.010
0.995	2.5758	0.0145	0.005
0.996	2.6521	0.0118	0.004
0.997	2.7478	0.0091	0.003
0.998	2.8782	0.0063	0.002
0.999	3.0902	0.0034	0.001
0.9995	3.2905	0.0018	0.0005

Bibliography

Alwin, D.F. (2007). *Margins of Error: A Study of Reliability in Survey Measurement*. Hoboken, NJ: Wiley.

Anastasi, A. (1982). *Psychological Testing*, 5e. New York: Macmillan.

Bagdonavicius, V., Kruopis, J., and Nikulin, M.S. (2011). *Non-Parametric Tests for Complete Data*. London & Hoboken, NJ: ISTE & Wiley. ISBN: 978-1-84821-269-5.

van den Berg, G. (1991). *Choosing An Analysis Method*. Leiden: DSWO Press.

Bland, J.M. and Bland, D.G. (1994). Statistics notes: one and two sided tests of significance. *BMJ* 309 (6949): 248.

Bryman, A. and Cramer, D. (2011). *Quantitative Data Analysis with IBM SPSS 17, 18 and 19: A Guide for Social Scientists*. New York: Routledge. ISBN: 978-0-415-57918-6.

Chrisman, N.R. (1998). Rethinking levels of measurement for cartography. *Cartography and Geographic Information Science* 25 (4): 231–242.

Cohen, J. (1983). The cost of dichotomization. *Applied Psychological Measurement* 7: 249–253.

Conover, W.J. (1999). "The Sign Test", *Practical Nonparametric Statistics*, 3e, Chapter 3.4, 157–176. Wiley. ISBN: 0-471-16068-7.

Corder, G.W. and Foreman, D.I. (2009a). *Nonparametric Statistics for Non-Statisticians: A Step-by-Step Approach*. Hoboken, NJ: Wiley. ISBN: 978-1-118-84031-3.

Corder, G.W. and Foreman, D.I. (2009b). *Nonparametric Statistics for Non-Statisticians: A Step-by-Step Approach*. Wiley. ISBN: 978-0-470-45461-9.

Corder, G.W. and Foreman, D.I. (2014). *Nonparametric Statistics: A Step-by-Step Approach*. Wiley. ISBN: 978-1-118-84031-3.

Cortina, J.M. (1993). What is coefficient alpha? An examination of theory and applications. *Journal of Applied Psychology* 78 (1): 98–104.

Cox, D.R. (2006). *Principles of Statistical Inference*. Cambridge University Press. ISBN: 978-0-521-68567-2.

Cronbach, L.J. (1951). Coefficient alpha and the internal structure of tests. *Psychometrika* 16 (3): 297–334. https://doi.org/10.1007/bf02310555.

Testing Statistical Assumptions in Research, First Edition. J. P. Verma and Abdel-Salam G. Abdel-Salam.
© 2019 John Wiley & Sons, Inc. Published 2019 by John Wiley & Sons, Inc.
Companion Website: www.wiley.com/go/Verma/Testing_Statistical_Assumptions_Research

Davidshofer, K.R. and Murphy, C.O. (2005). *Psychological Testing: Principles and Applications*, 6e. Upper Saddle River, NJ: Pearson/Prentice Hall. ISBN: 0-13-189172-3.

DeCastellarnau, A. and Saris, W.E. (2014). A simple procedure to correct for measurement errors in survey research. European Social Survey Education Net (ESS EduNet). http://essedunet.nsd.uib.no/cms/topics/measurement (accessed 19 October 2018).

DeVellis, R.F. (2012). *Scale Development: Theory and Applications*, 109–110. Los Angeles, CA: Sage.

Eisinga, R., Te Grotenhuis, M., and Pelzer, B. (2012). The reliability of a two-item scale: Pearson, Cronbach or Spearman-Brown? *International Journal of Public Health* 58 (4): 637–642. https://doi.org/10.1007/s00038-012-0416-3.

Essays, UK (2017). Assumptions, research design and data collection strategies. https://www.ukessays.com/essays/general-studies/underpinning-assumptions.php?vref=1

Fisher, R.A. (1925, 1925). *Statistical Methods for Research Workers*, 43. Edinburgh: Oliver and Boyd.

Freedman, D. (2000). *Statistical Models: Theory and Practice*. Cambridge University Press. ISBN: 978-0-521-67105-7.

Freeman, F.S. (1953). *Theory & Practice of Psychological Testing**. Sir Isaac Pitman & Sons, London.

Gault, R.H. (1907). A history of the questionnaire method of research in psychology. *Research in Psychology* 14 (3): 366–383. https://doi.org/10.1080/08919402.1907.10532551.

Geisser, S. and Johnson, W.M. (2006). *Modes of Parametric Statistical Inference*. Wiley. ISBN: 978-0-471-66726-1.

George, D. and Mallery, P. (2003). *SPSS for Windows Step by Step: A Simple Guide and Reference. 11.0 Update*, 4e. Boston, MA: Allyn & Bacon.

Gibbons, J.D. and Chakraborti, S. (2003). *Nonparametric Statistical Inference*, 4e. CRC Press. ISBN: 0-8247-4052-1.

Gibbons, J.D. and Chakraborti, S. (2011). Nonparametric statistical inference. In: *International Encyclopedia of Statistical Science*, 977–979. Berlin, Heidelberg: Springer-Verlag.

Grotenhuis, M. and Visscher, C. (2014). *How to Use SPSS Syntax: An Overview of Common Commands*. Thousand Oaks, CA: Sage.

de Gruijter, J., Brus, D., Bierkens, M., and Knotters, M. (2006). *Sampling for Natural Resource Monitoring*. Springer-Verlag.

Hand, D.J. (2004). *Measurement Theory and Practice: The World Through Quantification*. London: Arnold.

Harlow, L.L., Mulaik, S.A., and Steiger, J.H. (eds.) (1997). *What If There Were No Significance Tests?* Lawrence Erlbaum Associates. ISBN: 978-0-8058-2634-0.

Harmon, M. (2011). *Normality Testing in Excel-The Excel Statistical Master*. Mark Harmon.

Hettmansperger, T.P. and McKean, J.W. (1998). *Robust Nonparametric Statistical Methods*, Kendall's Library of Statistics, 1e, vol. 5. London: Edward Arnold. ISBN: 9780340549377. MR 1604954. also ISBN: 0-471-19479-4.

Hollander, M., Wolfe, D.A., and Chicken, E. (2014). *Nonparametric Statistical Methods*. Wiley.

https://en.wikipedia.org/wiki/Cronbach%27s_alpha.

IBM SPSS Statistics Base 22 (n.d.). IBM support. Retrieved from https://www-01 .ibm.com/support (accessed 23 Ocotober 2018).

JMP (2004). How do I interpret the Shapiro-Wilk test for normality? Retrieved March 24, 2012.

Joakim, E. (2011). *The Phi-Coefficient, The Tetrachoric Correlation Coefficient, and the Pearson-Yule Debate*. Department of Statistics, UCLA Retrieved from https://escholarship.org/uc/item/7qp4604r.

Kaplan, R.M. and Saccuzzo, D.P. (2009). *Psychological Testing: Principles, Applications, and Issues*. Belmont, CA: Wadsworth.

Kerby, D.S. (2014). The simple difference formula: an approach to teaching nonparametric correlation. *Comprehensive Psychology* 3: Article 1. https://doi .org/10.2466/11.IT.3.1.

Kline, P. (2000). *The Handbook of Psychological Testing*, 2e, 13. London: Routledge.

Kock, N. (2015). One-tailed or two-tailed P values in PLS-SEM? *International Journal of e-Collaboration* 11 (2): 1–7.

Kruskal, W. (1988). Miracles and statistics: the casual assumption of independence (ASA Presidential address). *Journal of the American Statistical Association.* 83 (404): 929–940. https://doi.org/10.2307/2290117. JSTOR 2290117.

Kuder, G.F. and Richardson, M.W. (1937). The theory of the estimation of test reliability. *Psychometrika* 2 (3): 151–160.

Laerd Statistics (n.d.). SPSS tutorials, nonparametric tests. Retrieved from https:// statistics.laerd.com/ (accessed 23 Ocotober 2018).

Leech, N.L., Barrett, K.C., and Morgan, G.A. (2005). *SPSS for Intermediate Statistics: Use and Interpretation*. Psychology Press.

Lehmann, E.L. (1997). Testing statistical hypotheses: the story of a book. *Statistical Science* 12 (1): 48–52. https://doi.org/10.1214/ss/1029963261.

Lehmann, E.L. (2006). *Nonparametrics: Statistical Methods Based on Ranks*. With the special assistance of H.J.M. D'Abrera (Reprinting of 1988 revision of 1975 Holden-Day ed.), xvi+463. New York: Springer. ISBN: 978-0-387-35212-1. MR 0395032.

Lehmann, E.L. and Romano, J.P. (2005). *Testing Statistical Hypotheses*, 3e. New York: Springer. ISBN: 0-387-98864-5.

Lenhard, J. (2006). Models and statistical inference: the controversy between Fisher and Neyman-Pearson. *The British Journal for the Philosophy of Science* 57: 69–91. https://doi.org/10.1093/bjps/axi152.

Levesque, R. (2007). *SPSS Programming and Data Management: A Guide for SPSS and SAS Users*, 4e. Chicago, IL: SPSS Inc. ISBN: 1-56827-390-8.

McCallum, R.C., Zhang, S., Preacher, K.J., and Rucker, D.D. (2002). On the practice of dichotomization of quantitative variables. *Psychological Methods* 7: 19–40.

McPherson, G. (1990). *Statistics in Scientific Investigation: Its Basis, Application and Interpretation*. Springer-Verlag. ISBN: 0-387-97137-8.

Mellenbergh, G.J. (2008). Tests and questionnaires: construction and administration. In: *Advising on Research Methods: A Consultant's Companion*, Chapter 10 (ed. H.J. Adèr and G.J. Mellenbergh) (with contributions by D.J. Hand), 211–236. Huizen, The Netherlands: Johannes van Kessel Publishing.

Moser, C.A. and Kalton, G. (1972). *Survey Methods in Social Investigation*, 2e. United States of America: Basic Books. ISBN: 9780465083404.

Mosteller, F. and Tukey, J.W. (1977). *Data Analysis and Regression*. Boston, MA: Addison-Wesley.

Nuzzo, R. (2014). Scientific method: statistical errors. *Nature* 506 (7487): 150–152.

Osborne, J.W. (2013). Is data cleaning and the testing of assumptions relevant in the 21st century? *Frontiers in Psychology* 4: 370. https://doi.org/10.3389/fpsyg .2013.00370.

Pillemer, D.B. (1991). One-versus two-tailed hypothesis tests in contemporary educational research. *Educational Researcher* 20 (9): 13–17.

Razali, N.M. and Wah, Y.B. (2011). Power comparisons of Shapiro-Wilk, Kolmogorov-Smirnov, Lilliefors and Anderson-Darling tests. *Journal of Statistical Modeling and Analytics* 2 (1): 21–33.

Ritter, N.L. (2010a). Understanding a Widely misunderstood statistic: Cronbach's alpha. Paper presented at Southwestern Educational Research Association (SERA) Conference 2010, New Orleans, LA (ED526237).

Ritter, N.L. (2010b). Understanding a Widely misunderstood statistic: Cronbach's alpha. Annual Meeting of the Southwest Educational Research Association, New Orleans, LA.

Robinson, M.A. (2018). Using multi-item psychometric scales for research and practice in human resource management. *Human Resource Management* 57 (3): 739–750. https://doi.org/10.1002/hrm.21852.

Royston, P. (1992). Approximating the Shapiro-Wilk W-test for non-normality. *Statistics and Computing* 2 (3): 117–119. https://doi.org/10.1007/BF01891203.

Saris, W.E. and Gallhofer, I.N. (2014). *Design, Evaluation and Analysis of Questionnaires for Survey Research*, 2e. Hoboken, NJ: Wiley.

Saris, W.E. and Revilla, M. (2015). Correction for measurement errors in survey research: necessary and possible. *Social Indicators Research* 127: 1005–1020. https://doi.org/10.1007/s11205-015-1002-x.

Sheskin, D.J. (2003). *Handbook of Parametric and Nonparametric Statistical Procedures*. CRC Press. ISBN: 1-58488-440-1.

Smith, T.W., Davern, M., Freese, J., and Hout, M. (1972–2016). General Social Surveys [machine-readable data file] /Principal Investigator, Tom W. Smith; Co-Principal Investigator, Michael Davern; Co-Principal Investigator, Jeremy

Freese; Co-Principal Investigator, Michael Hout; Sponsored by National Science Foundation. –NORC ed.– Chicago: NORC at the University of Chicago [producer]; Storrs, CT: The Roper Center for Public Opinion Research, University of Connecticut [distributor], 2017. 1 data file (57,061 logical records) + 1 codebook (3,567p.). – (National Data Program for the Social Sciences, No. 22).

Sprent, P. (1989). *Applied Nonparametric Statistical Methods*, 2e. Chapman & Hall. ISBN: 0-412-44980-3.

Swed, F.S. and Eisenhart, C. (1943). Tables for testing randomness of grouping in a sequence of alternatives. *Annals of Mathematical Statistics* 14: 83–86.

Tavakol, M. and Dennick, R. (2011). Making sense of Cronbach's alpha. *International Journal of Medical Education* 2: 53–55.

Triola, M. (2001). *Elementary Statistics*, 8e, 388. Boston, MA: Addison-Wesley. ISBN: 0-201-61477-4.

Triola, M.F. (2006). *Elementary Statistics*, 10e. Boston, MA: Pearson Addison Wesley.

Verma, J.P. (2013). *Data Analysis in Management with SPSS Software*. India: Springer.

Verma, J.P. (2014). *Statistics for Exercise Science and Health with Microsoft Office Excel*, 713. New York: Wiley.

Verma, J.P. (2015). *Repeated Measures Design for Empirical Researchers*. New York: Wiley.

Verma, J.P. (2016). *Sports Research with Analytical Solution Using SPSS*. New York: Wiley.

Verma, J.P. (2017). *Determination of Sample Size and Power Analysis with G*Power Software*. Kindle: Independently Published.

Wasserman, L. (2007). *All of Nonparametric Statistics*. Springer. ISBN: 0-387-25145-6.

Wilcoxon, F. (1950). Individual comparisons by ranking methods. *Biom. Bull.* 1 (6): 80–83. https://doi.org/10.2307/3001968.

Index

Testing Statistical Assumptions in Research, First Edition. J. P. Verma and Abdel-Salam G. Abdel-Salam.
© 2019 John Wiley & Sons, Inc. Published 2019 by John Wiley & Sons, Inc.
Companion Website: www.wiley.com/go/Verma/Testing_Statistical_Assumptions_Research